MISSION TO ABISKO

■

STORIES AND MYTHS IN THE CREATION OF SCIENTIFIC "TRUTH"

EDITED BY

JOHN L. CASTI

AND

ANDERS KARLQVIST

HELIX BOOKS

PERSEUS BOOKS

Reading, Massachusetts

ISBN 0-7382-0002-6

Library of Congress Catalog Card Number: 98-87902

Perseus Books is a member of the Perseus Books Group

Jacket design by Suzanne Heiser
Text design by John L. Casti

1 2 3 4 5 6 7 8 9 10—0201009998
First printing, December 1998

Find Helix Books on the World Wide Web at
http://www.aw.com/gb/

Contents

CHAPTER 9

CHAPTER 10

CHAPTER 11

MISSION TO ABISKO

Preface

In May 1997, during the period of the midnight sun, a small group of scientists, science-fiction writers and historians of science gathered to examine the stories that scientists tell—both to themselves and to the public at large—that serve to create what we may euphemistically call the "world view of science."

As is manifestly evident by even a casual perusal of the science section at a general bookshop, scientists are busier than ever telling stories about their craft for public consumption. What is far less evident, at least on the surface, is that scientists create even more elaborate tales that they tell to each other. These tales, often embodied in thought experiments like the famous alive/dead cat of Schrödinger or the infinitely large hotel of Hilbert, serve as ways to focus and encapsulate large amounts of knowledge in particular areas into short, pithy verbal pictures that capture the essence—and the shortcomings—of a scientific theory.

The writer of what's come to be termed "hard" science fiction is somehow caught between these two worlds. On the one hand, the writer must adhere to the stories scientists tell each other in order to be true to the cutting-edge science of today (and often tomorrow) that forms the underpinning to his or her tale. But a novel is a novel, not a scientific paper. This means that the sci-fi writer must also wrap the scientific story in a sugar-coating, so that the non-scientific reader can not only access the message, but also derive some entertainment value from it. Not an easy tightrope to walk!

The organizers of the Abisko seminar thought that perhaps these two groups of storytellers—the scientists and the science-fiction writers—might have something to say to each other. So the deal was done and a handful of writers with scientific credentials and scientists with writing in their blood got together for a week in the cozy confines of the Abisko Scientific Research Station to compare notes. Although no written record can begin to capture the full flavor of the interchanges that took place over the meals, coffee breaks, seminars, walks and other outings during that week, this book is a pretty good failure.

Acknowledgement

Beginning in 1983, the Swedish Council for Planning and Coordination of Research has organized an annual workshop devoted to some aspect of complex systems. These workshops are held in the scientific research station of the Royal Swedish Academy of Sciences in Abisko, a rather remote village of about 100 people located far above the Arctic Circle in northern Sweden. The editors and contributors wish to acknowledge the help of the Swedish organizers in making this meeting such a success.

John Casti, Santa Fe
Anders Karlqvist, Stockholm

MISSION TO ABISKO

THE ANALOGY OF NATURE

JOHN D. BARROW

The popularization of science has become a genre of our times: a bridge between the incomprehensible world of science and the comfortable familiarity of the arts. Its history is long. Not least among the reasons for the displeasure breathed upon Galileo was his dangerous habit of divulging his revolutionary discoveries to the educated general public in the vernacular. In Newton's time, popularization in England began to have an ulterior motive. The unveiling of the laws of Nature to reveal the universality, simplicity and harmony behind the cacophony of the appearances was presented as the ultimate proof of the design of Nature, and hence of a Grand Designer behind the scenes. Whilst few had read Newton, many had read about him, and there sprang up a crop of eccentric popular books presenting "Newtonian" theories of just about everything, like Desaguliers' *Newtonian System of the World, the best model of government*. The most successful was probably Count Francesco Algarotti's *Sir Isaac Newton's Philosophy for the Use of the Ladies*, which was not quite as condescending as it sounds. It was something of an enlightened novelty to present physics and mathematics to such an audience.

In recent years, scientists in the U.K. have been under attack from commentators, like Brian Appleyard, who see the exposition of new ideas in science as undermining the fabric of society and human values. They have argued that modern science creates an environment where certainty is indefensible, conservatism is eroded, and a demystified humanity is irrevocably diminished. In response, it is argued, apologetically, that the Universe is not constructed for our convenience. We just have to make the best of it. Humanity displays its moral and ethical qualities by the responses it makes to the world, not by suppressing the truth about it because it might be socially unhealthy.

The traditional guardians of the debate about "meaning of life" issues perhaps feel threatened by the entry of new scientific concepts into many of the areas that used to be their sole domain for voicing what could be merely opinions. Faced with being excluded by lack of familiarity with the agenda of new ideas and issues, a minority have over-reacted against the entire scientific enterprise. The fan of science and its popularization, on the other hand, sees it as the only real news. Everything else is a variation on a familiar theme of human relationships, politics, war and peace. The danger here is to get carried away with the immediate consequences of new discoveries, so that their importance is exaggerated and everything that follows is inevitably devalued. One has only to review the past publicity surrounding the discovery of temperature variations in the cosmic background radiation by the COBE satellite, or life on Mars, and the ensuing backlash, to appreciate this point. Scientific knowledge is not merely a fashionable flavor of the month.

Scientific popularization in English has inherited an interesting tradition, at least in its efforts to communicate progress in the harder physical sciences. If we sample a typical account of the frontiers of particle physics, or of astrophysics, we repeatedly encounter a familiar pedagogic device: the use of analogy. The confinement of quarks is "like" marbles inside a rubber bag; superstrings are like elastic bands; the quantum wave function is like a crime wave; a pulsar is like a lighthouse, and so on. This approach to the popularization of science has a distinctly Victorian air about it. It reduces the world to a collection images of practical devices familiar from everyday life. And that giant of Victorian science, Lord Kelvin, made constant use of this approach, not only in his explanatory writings, but in his research work as well. Interestingly, there were objectors to this style of thinking. Some continental mathematical physicists of the time found it exasperating that physicists like Kelvin were never content with a mathematical account of any natural phenomenon until they could translate it into an analogical picture involving little wheels or rolling balls. Whilst they were content with the abstract mathematical explanation, a Kelvin or a Faraday was not.

An interesting contemporary sidelight upon this state of affairs is provided by Stephen Hawking's staggeringly successful book *A Brief History of Time*. There is one area of this book which seems to stump many general readers, and ought to stump experts only they are so used to it that they pass hurriedly by, unperturbed. The point in question is a radical hypothesis proposed by Hawking and James Hartle for the initial quantum state of the Universe, the so called "no boundary condition." The key ingredient in their proposal is the requirement that time becomes another dimension of space in the quantum gravitational environment provided in the first moments of the Big Bang. This has all manner of interesting ramifications, some of which Hawking's book goes on to elaborate. We do not need to repeat them here. Rather, our interest is in this tantalizing statement that *time becomes another dimension of space*. We know what every word in this statement means. It makes good sense in ordinary English in a way that makes us think that we understand its content. But is any meaningful information being conveyed to the lay reader? The answer is probably "no." And the reason is interesting. Surely, it is because one here encounters a scientific idea which, as yet, seems to admit no handy everyday analogy. One can't say time becomes space is like a chrysalis becoming a butterfly, or even like a frog becoming a Prince. It's more like "the sound of one hand clapping." It is just a bare statement of the mathematical reality.

This is the first occasion that this impasse has been encountered in my scientific experience. It reveals why it is so difficult to get across the essence of this quantum cosmological picture of time at a popular level and why so many general readers found this idea a sticking point of the book. Perhaps it is evidence that the boundary we have reached here is not so much cosmological as akin to that suggested by Einstein's dictum that things should be made as simple as possible, but no simpler.

The dilemma of the missing analogy may have a deeper significance. Generally speaking, one can take two views of the activities of physicists who are seeking what they call "fundamental laws of Nature." Either you believe, as they often do, that they are discovering the real thing, and they will one day unveil the mathematical form of the ultimate laws of Nature. Alternatively, one may be more modest, and regard the scientific enterprise as an editorial process in which we are constantly refining and updating our picture of reality, using images and approximations that seem best fitted to the process. The latter perspective may, of course, still ultimately converge upon the former one in the course of time. An interesting feature of the latter view is that it links the popularization of science, and the analogical devices used therein, to the activity of scientific research itself. The constant updating of our scientific picture of how the Universe works can be viewed as a search for more and more sophisticated analogies for the true state of affairs. As they become more refined, they break down less and less often but, when sufficiently diluted, they become the homely images of the scientific popularizer. Viewed in this light, readers of popular science can see their activity as part of a continuous spectrum of imagery designed to make plain the workings of the world. They are closer to being actors than spectators.

The absence of a ready analogy for the central notions of new theories about the nature of time should not be viewed as a defect of those theories. On the contrary, it may be a healthy sign. Richard Feynman criticized some developments in fundamental physics for the enthusiasm with which they identified similar concepts and inter-relationships existing in superficially quite different areas of physics. Whereas some welcomed these cheap analogies as evidence of the deep unity of Nature, Feynman preferred to see them as a sign of our impoverished imaginations. We just couldn't think up any new structures! He suggests that it is just that "when we see a new phenomenon we try to fit it into the framework we already have — until we have made enough experiments, we don't know that it doesn't work. So when some fool physicist gives a lecture at UCLA in 1983 and says, 'This is the way it works, and look how wonderfully similar the theories are,' its not because Nature is really similar; it's because the physicists have only been able to think of the same damn thing, over and over again". So, when analogies from everyday experience are not readily available to illustrate new ideas in fundamental science, it may be a good sign that we are touching some brute fact of reality that cares not one whit whether there exist analogues elsewhere in the world. When there are simple everyday analogies of what goes on in the first exotic quantum moments of the Big Bang, the skeptic might be rightly suspicious that we are importing our existing ideas rather than discovering the new ones we need if we are to understand an aspect of reality that natural selection has not educated us about. Freeman Dyson once suggested that the trouble with the majority of hare-brained theories of elementary particles is not that they are crazy but that they are not crazy enough. True novelty and originality requires new images to be conceived, not merely the reprocessing of old ones. Perhaps the "time becomes space" mantra is just radical enough.

Scientific explanation has come to be equated with mathematical explanation. Social scientists and other "consumers" of mathematics are happy to regard it as a tool of human devising that is useful. But, for the fundamental physicist, mathematics is something that is altogether more persuasive. The farther one goes from everyday experience and the local world, the correct apprehension of which is a prerequisite for our evolution and survival, the more impressively mathematics works. In the inner space of elementary particles or the outer space of astronomy, the predictions of mathematics are almost unreasonably accurate. If one takes matter to pieces and probes to the root of what those pieces "are" then ultimately we can say nothing more than that they are mathematics: they are relationships. Moreover it is not just the quantity of mathematics that imbues the structure of reality that is impressive, it is its quality. The sort of mathematics that lies at the heart of the general theory of relativity or of elementary particle physics is deep and difficult, far removed from the lay-person's conception of mathematics as a form of high-level accountancy. This has persuaded many physicists that the view that mathematics is simply a cultural creation is a woefully inadequate explanation of its existence and effectiveness in describing the world. If mathematics is discovered rather than invented—if "pi" really is in the sky—then we can say something more about the analogical structure of the world. For, when we see the continuing refinement of our image of Nature in the development of more abstract mathematical theories of her workings, then asymptotically we are learning something about mathematics. If mathematics is just another human idiom which captures some but not all aspects of the world, then it is yet another analogy that ultimately fails. But if the world is mathematical at its deepest level, then mathematics is the analogy that never breaks down.

A curiosity in this respect has been the investigation of the theory of superstrings. After early euphoria at the special nature of these "Theories of Everything," the sober reality of the situation has emerged. Laws of Nature are much simpler than the outcomes of those laws that make manifest the world around us. This is a reflection of the fact that mathematical equations are much easier to find by appeal to general symmetry principles than they are to solve. Whilst we may be in possession of the mathematical theory of superstrings, it is beyond our present mathematical capabilities to extract the predictions and explanations of the world that the theory has to offer. One of its foremost investigators, Princeton physicist Ed Witten, has remarked that we may simply have been lucky enough to stumble upon this theory fifty years too early: before mathematics was sufficiently advanced to handle it. "Off the shelf" mathematics has not been sufficient to unveil the inner secrets of superstrings. And, in fact, the quest for those secrets has pointed mathematics in new and fruitful directions of great novelty. Again, this lack of progress with a theory of physics may be a healthy sign that we have taken the hard path that leads to the truth rather than the easy way of wishful thinking that seeks progress for its own sake. But while superstring theory has its novel aspects—allowing additional dimensions of space and an abandonment of the traditional picture of the most elementary constituents of matter being point-like—is it radical enough? Maybe, but if only superstrings weren't so much like little rubber bands.

2

PROVING THE DREAM

GREG BEAR

Niels Bohr once said "It is wrong to think that the task of physics is to find out how nature is. Physics concerns what we can say about nature." In a way, we are all writing stories about the world, but some of our stories seem to bear a mysterious relationship to the world that others do not. They use a particular syntax, even a different grammar, and they follow certain rules. These rules give them a resonance with the real, and the stories acquire real usefulness. But just behind them lie stories which bend the rules, do not have an immediate power; these are the fictions that lie behind the more powerful stories that lie closer to fact. All are made of the only real tool our mind has, language.

We observe – and turn light into neural language, storing it away as chemical compounds which will later, perhaps, influence our actions. We try to communicate what we observe, expressing ourselves in another kind of language, written or spoken or through other signs. These texts influence other individuals; they are like the neurotransmitters that convey information and commands between neurons. Some of the texts are especially useful and immediately powerful, others have latent functions that seem to bloom and affect our lives in very different ways.

I once asked Stephen Hawking (who has never, to my knowledge, written fiction) whether his best ideas began as poetic insights. "Yes," he said. "But then I have to make them work." It's not that much different with fiction, though the tools used appear at first sight to be very different. Telling a story is an explanatory act, much as working up a theory seeks to explain. Explanation is common to both science and fiction. It is no surprise that many scientists try their hands at writing fiction; it's either a common malady or it's a clue as to how we all work, in our heads, when the social masks are off and nobody's watching and the private labor of explanation and creation begin.

Science and mathematics make up the grammar of the language we use in modern Western culture to order our priorities and find the best and most efficient ways to get our work done. They have been distilled from thousands of years of personal experience; I suspect the mindset of the first flintknapper was little different from that of a modern laboratory technician.

But what about the grand schemer: the theoretician, whose work has always been protected from practical applications by a wall of almost religious presumption? Was there always a master flint theoretician, who knew where the flint nodules could be found, and why they were there, and not somewhere else, and what might be done to improve hand axes? Here's a very important and useful individual who quite probably wouldn't have been caught dead knapping. He (or perhaps she — will we ever know?) needed long hours of contemplation, of observation and being left alone. Perhaps this master of flint had a religious vision that ordered things and made it easier to understand the mysteries of flint. And then some chucklehead saw a seashell embedded in the flint and had a different vision, which he tried to promulgate among the tribe, perhaps using the mutterings of the master of flints as a reference and authority. This could have been the first science fiction writer. Or the first cosmologist, who knows? I often have a hard time telling them apart. At any rate, my scenario could explain the poor victims of skull bashing seen at certain archaeological sites. They were cosmologists and science fiction writers who disappointed their audience.

A story is an attempt to order a set of events, real or imagined, to place those events in a sequence and package them in such a way that others will want to listen to them or read them. An hypothesis is an attempt to order a set of observations, to package them in such a way that they will be taken up by one's peers and used to explain later observations. A confirmed hypothesis becomes a theory. A confirmed story is a myth or a best-seller. A good theory gives us both explanation and power. A good story gives us the feeling of understanding events, which is a kind of power, and of course a kind of explanation, in a cultural context.

A good theory, however, can be very dangerous. A good story more often than not seems fairly innocuous.

Some say that the difference between science and art is that science is supposed to rise above culture and personality. Of course, it does no such thing. Science and art are both social and cultural acts. Scientists are human — at least as human as science fiction writers — and the way they think is often embedded in and supportive of their culture. What we most admire about science, however, is the ability it often gives us to rise above our incorrect preconceptions, our blindnesses, to make a set of observations, and order them in such a way that we have a useful insight no one has ever had before. This insight, rather than being shaped by culture, now turns around and shapes the culture. The individual who has this insight is regarded with some awe. Science too creates a myth, but it is a myth that engineers find compelling; a myth that lets bridges be built and not fall down. Other myths seem to pale by comparison.

Science fiction is story-telling in the context of a scientific culture. All contemporary fiction, to be truly contemporary, must involve some scientific content. Science fiction, however, adds a different challenge. What will science and technology and culture be like *in the future?* A really startling science fiction story can also rise above the culture and shape it. The individual who writes this story is regarded with some disdain. They are not always, after all, scientists, and many critics doubt that they are really story-tellers. What are they? Scientific want-to-bes?

There's a certain truth to this accusation. I had a fairly heavy education in science and math, and made short stabbing thrusts at the target of a scientific career. But what I've always truly wanted to be is what I am — a science fiction writer. If I could have been a scientist, too, that would have been fine, but it did not turn out that way.

Anyone who studies science in our scientific culture soon notices (and quite often just as soon ignores) a curious relationship. The first urge to do science is often instilled by a youthful reading of science fiction. Perhaps policemen decide to get into law enforcement after reading mysteries — I don't know — but most of the scientists I have met in nearly all branches of science, read science fiction in their youth. Many continue to read it—and I might point out that these are very highly regarded scientists, among the best. (Years ago, the story got around in the newspapers that Linus Pauling came up with his alpha helix model of protein structure while lying in bed recovering from a cold. The newspapers said he was reading mysteries. Only later did we learn that he was actually reading both science fiction and mysteries! But for a scientist to read science fiction was not then regarded as respectable. And if they wrote it, they often used pseudonyms. Fred Hoyle was a conspicuous exception!)

Science fiction is fun, and at its best, it doesn't insult one's intelligence. It often has scientists as characters—a rather flattering attraction—and sometimes it incorporates interesting speculations about science. On rare occasions these speculations are so interesting they point the way to new developments.

In a way, science fiction writers dabble in science, suggesting interesting scenarios that incorporate intriguing hypotheses. Then we go away and let the scientists obtain the grants and do the work, and hope to get some of the credit. Science fiction writers and scientists are in both symbiotic and parasitic relationships. Sometimes we co-venture, sometimes we steal ideas without credit—but always, we cross-pollinate. What science fiction does best of all is model the results of scientific progress. The consequences of science are expressed in story. The scientist might then read science fiction as a kind of cultural cross-checking, to see just how things might change given this or that development. The science fiction writer in turn keeps track of scientific developments in the hopes of staying current and remaining convincing a little while longer.

In the long run, of course, neither scientists nor science fiction writers have perfect track records in prediction. We are neither of us very good at prophecy. But I think it can be shown that science-fiction writers have an edge over the scientists. We are often less dragged down by having to pay our respects to a peer-review process. We can be a little more wild and irresponsible. We don't have to keep our grants funded—we only have to fulfill contracts, a less complicated if no less aggravating process. This looseness more closely approximates the informal chaos of reality. Hence, our future scenarios have a certain life to them that straightforward, by-the-numbers prognostications do not.

Perhaps this is why many of the best scientists write science fiction at some time or another—it's part of a natural curiosity about how things work or will work. The story is a kind of machine that helps us think differently about our discoveries. It automatically places knowledge and theory in a social context— that is, in a context that involves characters and social situations, expressed through time in a complex world.

In the beginning, a scientific theorist tries to find the roots of a story that may have already played out in a laboratory or in nature. Scientists try at first to remove those human variables that obscure the facts: wishful thinking, hope for career advancement, loyalty to old theories or teachers or colleagues, and so on. Once these impediments are removed, the work proceeds more efficiently, and the insights are more likely to be useful. This is the power of the scientific method. The story we are trying to uncover is not our personal story: it is the story of the phenomenon we hope to explain. Once that story becomes clear to an individual or group, however, it must be returned to a social and cultural context. It must be submitted to other scientists, chewed over, argued, confirmed or dismissed. The story becomes part of a greater story, and if it succeeds, if it survives and is useful, it becomes part of the lexicon of all scientists, part of the bag of tools available to any society or culture that cares to understand the world. This turning-into-tools of a theory actually shapes thought and language.

The lexicon of science implies a grammar of science. And science fiction writers, like scientists, are keenly aware that science has a kind of grammar. The ideas themselves fit a syntax; certain statements seem to make more sense, be more grammatical, than other statements. They just seem more scientific. Scientists struggle to fit their theories into this grammar, to make the theories fit within a larger conception, a greater explanation.

Some grammatical rules in science are in fact unproven assumptions, axioms of scientific culture, common tribal wisdom. Form follows function. The universe is isotropic. One law for all regions and all cultures. Others seem to express a deep structure of the universe at all levels: You can't get something for nothing. You can't break even. In fact, these bits of tribal wisdom share striking similarities with any practical discipline, as well as pragmatic philosophy. Engineers used laws long before they were

scientifically explained and codified, simply because they worked and were useful.

We can smell out scientific wannabes pretty easily. They smile funny when they talk. They stick to their facts even when the facts are demonstrably false. Their ends define their means and their discourse. And they tell us things that don't fit the scientific grammar: you can get something for nothing; different cultures see nature differently, so nature must follow different laws for different people; numbers are too precise to be reliable. A sucker is born every minute — and time's a-wasting.

Bad science fiction may tell good stories — entertaining stories — but it fails this grammatical test just as much as the suspect stories of modern charlatans and other scientific wannabes. Bad science fiction tells the public what it wants to hear, not what's more likely to be true.

The best science fiction does not cheat, any more than does the best science. But bad science fiction is not alone in cheating on the facts. Scientists of great reputation have been known to tell half-truths and out-and-out distortions to advance a cause. We are all of us human, and the road to bad story-telling is the same as the road to the politically expedient half-truth: whichever will get us the larger, more appreciative audience. That is the best route for some to follow. Now I'm going to plunge into the most rarefied realm — that of the visionary. It is here, I think, that the dividing lines between science fiction writer, scientist, and poet fade into invisibility. The truly new is not always useful, and it is certainly not obvious — art and intellect unite to explore new territory. Clues to the truly new vision can lie anywhere; the one thing that can be said about the visionary is that she goes where everyone else sees only darkness or debris or unassociated facts. How does the visionary see a pattern? By working from a structure that is at times ineffable, that is, above linguistic expression. The highest and simplest expression of this pattern is the visionary himself.

You can feel an original vision. It comes like a surge of ghostly hormones, tapping at your backbone. It can either raise the hairs on your neck with its dangerous strength, or it can whisper a suggestion, a few leading phrases. Either way, you are in trouble. An original vision can be wrong. It can lead you into a wild goose chase. But its strength is such that you must follow it. If you're

smart, you arrange a series of high intellectual hurdles for this vision to clear; it must be more than just new, it must provide deeper insight, lead to new avenues and more visions. If it circles back and traps you in the mundane and the obvious, it is not genius; it has the sickly phosphorescent glow of error. If it drags you down and confirms the prejudices of your culture, it's likely to be wrong as well, and probably not original.

The same is true for a truly startling poem or story. And they come from the same roots, though perhaps in different individuals! There have been scientists and writers good at one thing and brilliant at another, but very few brilliant at both science and art. Still, the poetic impulse is the beginning, and it arises from the brain-body, the very structure of the individual.

A human being is not autonomous. She is connected to a social body and can't survive for long without a little help from her friends. A human brain is likewise connected to a body, and the two have never been and perhaps never will be successfully separated without damage to both. The human body is a brain with muscles and motility, organs of sense and organs of communication. Brain and body work together; they think together. Anything else is cultural misconception, illusion. Thought arises as a complex of physiological changes, flushes of blood in body and brain tissues, providing energy for certain active and demanding subsystems. The expression—our conscious signaling of the act—comes later, first as an awareness of an impulse or solution, later perhaps as an expression to others, a report on the results of our physiological processes.

The culture in turn is a neural processing system, with humans as nodes in a social network, a diffuse, distributed body. The culture thinks over the results of the individual's thought-work, weighs it, finds it attractive or repulsive, useful or useless, provides more incentive if useful, less if deemed useless. Just as original visions can be stifled within an individual and never expressed, through some flaw or reticence, an original and useful vision may be stifled or suppressed within a culture.

Bodies both social and individual, like all neural networks, are prone to error. The survival strategy of a neural network is that it is designed to quickly handle many large and diffuse problems that do not submit to simple solutions, and it is more often right than wrong. If the cultural body has a physiology—

and I think it does—it may have an analogous endocrine system. A healthy culture encourages the expression of individuals, even though that expression may contradict cherished beliefs. This encouragement comes in the form of access to goods and services, as it were. Labor, work in any form, is not so much rewarded as fueled. In a body, the fuel is supplied directly through blood flow. In a culture, money is supplied, and money authorizes access to the labor of others, to either encourage work to be done for the individual, or to utilize goods already in existence. Money is a social hormone that authorizes the release of work or the consumption of resources.

One of the hallmarks of an original vision is that it will often meet strong resistance. A truly new idea may reorganize the social structure and cause immense effort. It must be submitted to extraordinary scrutiny and confirmation before it becomes part of the cultural anatomy, a directive or guide on a culture-wide scale.

For this reason, the originality of the artist is far less danger-ous than the originality of a politician of lawyer or scientist. The artist slips us visions that are "fanciful," not meant to be taken seriously. They do not signal social revolutions, but sanctioned play; whimsy, polite revolutions on a small scale (Tyrants have an instinctive eye for originality, and try to squash it. Tyrannies are cancerous organizations that empower a small group or an individual, at the expense of all other groups, and they typi-cally halt the overall growth of a culture, sometimes for decades. The suppression of new ideas is a danger sign of an impending tyranny.)

When the science fiction writer produces a startling new idea, a new story, it is regarded as play; but it also greases the skids for further development. Scientists mull over the possibilities, it becomes part of the subconscious current of potential change, a candidate for future exploration and release of resources. If the idea turns out to be ludicrous, no harm done! It was all a lark anyway.

The idea may be so wrapped up in symbols that it is totally ignored by a usually vigilant censorship committee; it will not be attacked because it is so obviously confusing and unimpor-tant. (Science fiction and fantasy have typically slipped by the tyrannical moral and political censors of modern cultures. They

simply don't understand; so they ignore. In a conversation with
Boris and Arkady Strugatsky, Russian writers of science fiction,
we encountered a serious disagreement. I stated that metaphor
was a diversified tool of many uses, weakened when given a
purely political function; they fervently believed that metaphor
was nothing if it was not used as a tool for criticism of the
state!) The culture is full of subconscious murmurings, dreams
and visions and playful intellectual baubles. They signify a healthy
ferment, rather like the streaming of an intellectual protoplasm.
But they also act as a store of possible solutions for future problems, reactions to future situations. It's been said that science
fiction immunizes us against future shock; it also works hard to
keep us from being completely surprised. And it performs this
task without committing too many valuable resources!

At the top of the pyramid, the visionary sees very far, but
perhaps he's looking in the wrong direction. No matter; he's a
crackpot anyway, best ignored.

When the visionary turns out to be seeing in the right direction, her ideas are immediately taken up by hundreds or perhaps
thousands of others, and made their own. Some claim the ideas
are original with them; perhaps they are. The larger the social
body, the more visionaries can be supported, the more originality
can be spread around.

Next, the ideas are submitted to the intellectual subsystems
of the culture, for further analysis. Nanotechnology is in this stage
now: not quite an industry, but not simply a vision, either. At
this stage (or even before), these ideas are spread throughout
the culture for general consumption, to get the culture ready
for an impending change. The vision now becomes cliché. It is
expressed in popular culture, in books and films; it becomes a
household concept, though it is not yet real. Robots are both
part of popular culture and part of industry, but the intellectual
robot, the artificial intelligence, is still more a cliché than a reality! Development has not yet matched the dream, the original
vision.

By the time the concept is demonstrated in application or
in actual discovery (cloning of complex mammals, possible life
on Mars, oceans on Europa, atom bombs, space travel, etc.) the
culture is prepared. Both science and science fiction have played
a necessary part; both interact and support each other. They add

words and additional refinements of grammar to the language of science, which is the language of our culture.

Someday, a tyrant is going to recognize the power of science fiction, and try to harness it. She will force all scientists to express their grant proposals in the easily digestible form of a good story. The story will be submitted for cultural approval, turned into a cliché, made into a series of expensive motion pictures full of glorious special effects, and then denounced as boring and unoriginal. The idea will then be extinguished and its ashes swept under the rug.

Until that happens, we are all in this together, collaborating, stealing, hogging the limelight, crying out for attention. Science fiction is the unproven dream of science. That so few people understand or condone this relationship is only healthy; we escape the ire of the censors, and get on with our revolutionary work, which is, after all, discovering what we can say about reality, and then changing the world.

3

BEYOND THIS HORIZON: ENVISIONING THE NEXT CENTURY

OR

STORIES OF OUR (PREVENTABLE?) FUTURE(S)

GREGORY BENFORD

Introduction

Much has been said about the inexorable link between society and science. I assume this coupling is inevitable and useful. More, I shall take a full engagement with our possible futures as a positive good, far better than hedgehog reluctance, which often masquerades as mere moral nostalgia.

We scientists who choose to speak to the rest of our society bear a particular burden. We cannot discuss where science may lead without using imagination, yet science itself is supposed to be strictly true in the checkable, provisional sense. In a sense, by discussing possibilities that may not come to pass, we are telling lies about the truth.

How, then, to prepare society for events which will dazzle and frighten? Many coming events will be conceptual bugs on our social windshield, seen too late to do anything about. To avoid this, some balanced, historically sophisticated way of viewing our increasingly technological future is needed.

Here Comes the Future, Ready or Not

As the millennium approaches like an overloaded freight train, fat with metaphor, we shall see many attempts to peer beyond the veil of that magical number, 2000. (For purists, 2001.) Many try to do linear extrapolations from current trends. Others assume, like southern Californian weather forecasters, that tomorrow will be pretty much like today, only more crowded.

Perhaps the best approach uses analogy. Could our century have been foretold in the 1890s?

First, recall that the nineteenth century was dominated by the metaphors and technological implication of two sciences: chemistry and mechanics. Wonders as striking as railroads and steamships conspired with humble revolutions like the use of artificial fertilizer to make the world new and bountiful.

To be sure, other themes were faintly sounding through the serene Victorian atmosphere. At mid-century the audacious Darwin-Wallace theory of evolution by natural selection began preparing the ground for modern biology, and excited enormous public furor. Elsewhere in England, Michael Faraday and James Clerk Maxwell were laying the foundations of electromagnetic technology. Their discoveries promised much that could be used fairly soon in practical ways. Darwin troubled minds, but had no useful implications.

While the older crafts and models of Newtonian mechanics and workaday chemistry drove the great economic and social engines of the Victorian era, in the waning decade of the century, Edison, Marconi and others sounded the opening theme of the next, electric era. These inventors caught the public's imagination. Radio saved the Titanic's passengers, Edison captured movement and sound. The great, unsettling conceptual shifts of relativity and quantum mechanics followed later. By 1910 the transition was obvious.

For clearly, physics has dominated our century. It has altered everything, from the A-bomb to the electric vibrator. Electromagnetic theory and experiment gave us the telephone, radio, TV, computers, and made the internal combustion engine practical— thus, the car and airplane, leading inevitably to the rocket and outer space.

The fateful wedding of that rocket with the other monumental product of physics, the nuclear bomb, led to the end of large-scale strategic warfare—as profound a change as any in modern times.

Even now, as the century wanes, physicists remain our scientific Brahmins. They dominate government committees, holding forth on topics far beyond their nominal expertise—defense, environmental riddles, social policy. Yet in our growing environmental problems and the rapid advances in other laboratories, far from the physics departments of the great campuses, a clarion call is sounding through our time. Biology has turned aggressively useful.

By analogy, we may stand on the threshold of the Biological Century. Like the 1890s, our decade bristles with striking biological inventions. Conceptual shifts will surely follow. Beyond 2000, the principal social, moral and economic issues will probably spring from biology's metaphors and approach, and from its cornucopia of technology. Bio-thinking will inform our world and shape our vision of ourselves.

The Easy Era

While the particle physicists desperately, unsuccessfully tried to get their Superconducting Super Collider built in Texas, against growing opposition to the $10 billion price tag, a smaller initiative quietly proceeded: the Human Genome Project. This vast effort, eventually costing about $3 billion, will map the human genetic code—our DNA.

The Project's first director was James Watson, co-discoverer of DNA with Francis Crick. It is the largest job ever attempted in biology, but surely not the last foray of biologists into Big Science, where physicists have their own plantations. Already a bacterium genome was completely sequenced

Such sequencing opens vast ethical issues. We shall be able to know who has defective genes. What will it mean when we can be sure we're not all born equal? Worked out, the implications will scare a lot of people. Insurance companies will not want to cover those with a genetic predisposition to illness, for example. Here lurk myriad lawsuits.

But these are short-term ethical questions, surely. The true solution lies in fixing genes, not merely reading them. If parents-to-be can have their problem genes edited into normal ones, most of the issues may evaporate.

And this is just one of many advances which portend much. Will we stop at cleaning up what we see as defects? I doubt it.

As we all saw in grade school, once you learn how to read a book, somebody is going to want to write one—that's how authors like me are made. Once we know how to read our own genetic code, someone is going to want to rewrite that "text," tinker with traits—play God, some would say.

True rewriting lies a few decades off, I believe. The first years of the Biological Century will probably be an Easy Era, much as physics enjoyed a period of largely uncritical acceptance of wonder after wonder, until The Bomb. (Remember that radium was widely thought to be a general cure, until Madam Curie died of her exposure.)

The first signs of a quiet revolution in our daily lives will probably come with some fairly noncontroversial commercial products. Much research has gone into cellular critturs which can digest oil spills or other toxic contaminants. Some work reasonably well already. Soon enough such research will give us a spectrum of organisms which digest unpleasant substances. That should mean refineries which don't stink, rivers that don't catch fire, streams that aren't sewers.

Plants have plenty of chemical defenses, and a smart farmer will come to use that. In temperate zones, winter is the best insecticide; it keeps the bugs in check. The tropics enjoy no such respite, so plants there have developed a wide range of alkaloids which kill off nosy insects and animals. Nicotine is an excellent insect foe; the fact that we addict to it is a curious side effect. Adapting such defenses to orchards and crops is an obvious path for biotech.

Consider the farm of the next century, which we might better call a "pharm"—because it may well be devoted to growing proteins, not wheat. Already researchers can synthesize proteins in animals by co-opting their own schemes for making, for example, milk.

Genetically altered goats have been made to yield in their milk a particular human protein which effectively dissolves the fibrin clots responsible for coronary occlusions. Efficiencies are low, but probably won't remain so. To get high yields, it will be a good idea to go to the dairy cow, which produces 10,000 liters of milk a year.

Imagine a cow which yields insulin, the expensive lifesaver of diabetics. We could make such a cow by editing its genes which control the cow's internal chemistry. The simple way would be to make two kinds of cows, one which produced milk rich in the "alpha" chain that helps make up insulin, and a second which makes the "beta" chain. This would free the cows from having to contend with insulin in their own systems, for only when the alpha and beta chains are mixed do we get insulin itself.

Insulin grown down on the pharm would probably be much cheaper than ours today. Similarly, there seems no barrier to making many pharmaceuticals in natural systems. Sheep might be specialized to a whole range of useful drugs, for example. International giants such as Merck are studying such avenues now.

Sheep, goats and cows would become the essential "bioreactors" which reproduce themselves in a barnyard biotech which could benefit many farmers who never heard of protein tinkering. But there will be troubles, because such animals don't breed true. A dairyman in Argentina will have to come back to Pharms Unlimited for his next calf. Indeed, Pharms Unlimited would be mad to make its cows so they can reproduce their (patented!) technology without a fat fee. So the Third World may see this as just another way to keep them on an unending economic string.

Such technology will spread into the immensely profitable realm of direct consumer goods. The easiest will be items so commonplace that on television they look like simple extensions of what we already have.

Imagine a kitchen cleanser which dissolves waste in those hard-to-get places, maybe even invading the grouting of tile in pursuit of fungus. Present cleaners like Tilex have to be sprayed on, wait and rinse. A living variety could patrol on its own, digging very deep, then be rolled up into its holder, lying dormant until needed again. You won't run out of it.

Or ponder a bath mat which slowly tugs itself across the floor, slurping up puddles, deposits of soap and hair spray, hairs, general "human dander." It lives on the stuff, plus an occasional helpful dollop of diet supplements from the otherwise distracted homemaker—who thinks of it as a rug, not a pet.

Many products—the opening wedge—will be less startling. Invisible, convenient, they'll come innocently, not seeming to announce a revolution. Resident "toothpaste" that does the essential policing up after lunch, and maybe even makes your breath smell, well, not so bad. Stomach guardians which ward off Montezuma's Revenge before you notice a single symptom—permanently, because the microbes are symbiotic with you, and live throughout your digestive system.

Even insulin cows may become obsolete if we develop microbes which can do the essential task directly in the body. We now can modulate some controlling genes which supervise elaborate biochemical transactions. It seems feasible within a decade or two to tailor these controller genes exquisitely. Then we could produce a capsule containing controller microbes which directly sense the patient's blood glucose level.

This scenario has been suggested by molecular microbiologist Mark Martin of Occidental College. Once inserted in the body, such a capsule would respond to glucose changes by making insulin, just as the patient's body should. With a specially designed capsule wall which lets nutrients and glucose in, and insulin out, but keeps the bacteria confined, side effects would be minimal. This could be far more reliable than having patients clumsily self-sample and inject, as they do now.

Such medical improvements face little opposition, particularly if they are hidden inside the body. Overt changes will not be so welcome. Some, though, could be attractive.

Market forces seem likely to spur imaginations. Fad blends easily with fantasy. Maybe there will be a fashion in bio-corduroy, which lives off your sloughed-off skin, perspiration—and even, if you like, some of your less agreeable excretions.

The theme here is biological balance—what's waste to one creature can become food to another—with a desirable job done in the transaction. This is "homeostasis," the biological equivalent of the thermostat. Such ecology on a small scale could become a public sign of trendy virtue, as popular as recycling is today.

Biotech revisits ancient arts. For many millennia we've been breeding cows and corn, collard greens and collies, to our whim.

We can expect more exotic foods, of course, but more important, we may see new and better ways of growing them.

Ants and Analogies

Consider a field of maize — corn, to Americans. At its edge a black swarm marches in orderly, incessant columns.

Ants, their long lines carrying a kernel of corn each. Others carry bits of husk; there an entire team coagulates around a chunk of a cob. The streams split, kernel-carriers trooping off to a ceramic tower, climbing a ramp and letting their burdens rattle down into a sunken vault. Each returns dutifully to the field. Another, thicker stream spreads into rivulets which leave their burdens of scrap at a series of neatly spaced anthills. Dun-colored domes with regularly spaced portals, for more workers.

These had once been leaf-cutter ants, content to slice up fodder for their own tribe. They still do, pulping the unneeded cobs and stalks and husks, growing fungus on the pulp deep in their warrens. They are tiny farmers in their own right. But biotech had genetically engineered them to harvest and sort first, processing corn right down to the kernels.

Other talents can be added. Acacia ants already defend their mother trees, weeding out nearby rival plants, attacking other insects which might feast on the acacias. Take that ability and splice it into the corn-harvesters, and you do not need pesticides, or the drudge human labor of clearing the groves. Can the acacia ants be wedded to corn? We don't know, but it does not seem an immense leap. Ants are closely related and multi-talented. Evolution seems to have given them a wide, adaptable range.

Following chemical cues, they seem the antithesis of clanky robots, though insects are actually tiny automatons engineered by evolution, the engine that favors fitness. Why not just co-opt their ingrained programming, then, at the genetic level, and harvest the mechanics from a compliant Nature?

Agriculture is the oldest biotech. But everything else will alter, too.

Mining is the last great, traditional industry to be touched by the modern. We still dig up crude ores, extract minerals with great heat or toxic chemicals, and in the act bring to the surface unwanted companion chemicals. All that suggests engineering must be rethought — but on what scale? "Biomining" is actually quite ancient. Romans working the Rio Tinto mine in Spain 2000 years ago noticed fluid runoff of the mine tailings was blue, suggesting dissolved copper salts. Evaporating this in pools gave them copper sheets.

The real work was done by a bacterium, Thiobacillus ferroxidans. It oxidizes copper sulfide, yielding acid and ferric ions, which in turn wash copper out of low grade ores. This process was rediscovered and understood in detail only in this century, with the first patent in 1958. A new smelter can cost a billion dollars. Dumping low quality ore into a sulfuric acid pond lets the microbes chew up the ore, with copper caught downhill in a basin; the sulfuric acid gets recycled, at trivial cost. Already a quarter of all copper in the world, from Peru to Alaska, comes from such bio-processing.

Gold enjoys a similar biological heritage. The latest scheme simply scatters bacteria cultures and fertilizers over open ore heaps, then picks grains out of the runoff. This raises gold recovery rates from 70 percent to 95 percent; there is not much room for improvement. Phosphates for agriculture can be made with a similar, two-bacterium method. By noticing that our mining waste was food for another phylum, we close a biological loop, cleanse our human-centered world of "pollution," extend our resources and make a profit.

All these developments use "natural biotech." Farming began by using wild wheat — a grass. Antibiotic therapy first started with unselected strains of Penicillum. We've learned much, mostly by trial and error, since then. The next generation of biomining bacteria are already emerging. A major problem with the natural strains is the heat they produce as they oxidize ore, which can get so high that it kills the bacteria.

To fix that, researchers did not go back to scratch in the lab. Instead, they searched deep-sea volcanic vents, and hot springs such as those in Yellowstone National Park. They reasoned that only truly tough bacteria could survive there, and indeed, found some which appear to do the mining job, but can take near-boiling temperatures.

Bacteria also die from heavy metal poisoning, just like us. To make biomining bugs impervious to mercury, arsenic and cadmium requires bioengineering, currently under way. One tries varieties of bugs with differing tolerances, then breeds the best to amplify the trait. This can only take you so far. After that, it may be necessary to splice DNA from one variety into that of another, forcibly wedding across species.

We are already in the Easy Era. Around us, often without fanfare, emerge new technologies: engineered drugs, pest-resistant plants, single-gene alterations in plants and animals, genetic diagnostics (usually only DNA testing of suspects makes the news). Within two decades we shall see "bioactive" products which work and live among us and in us, engineered organisms, "pharms," and limited genetic editing.

Alas, only in retrospect shall adjusting to these changes seem easy.

Cultural Convergence

From our blinkered perspective, a Biological Century looks like a fundamental shift in world view, able to ramify into every cultural corner.

Physics proceeds by atomizing nature, and this habit of mind has deeply penetrated realms far from science. Every writer knows that the trick in literature is immersing the reader in a world, a knitted vision—yet some schools of criticism have aped physics, deconstructing literature until it is a swarm of disjoint words, each ambiguous, their author irrelevant. This stress on contradictory or self-contained internal differences in texts—jam jars or novels alike—rather than their link to a culture of meaning, merely leads to literature seen as empty word games.

A biologically sophisticated world view would counter this, looking for how artists and writers manage their integrative effects. Current academe despises genres, from the western or the musical, which many feel are the true strengths of American culture. This may be because genres don't yield to theories of atomized arts, but are best seen as cross-talk, conversations within communities, progressing down through time—evolving. Genres clearly unfold and interact, ragtime to jazz to blues to rock, suggesting biological metaphors.

What would a criticism look like, done in a biological style? Both species and genres have intense interior interactions. Seen in the large, population statistics in literature and life may follow similar laws, though of course occasional individuals can deflect the prevailing tide, in both culture and in survival fitness.

Weighing these seemingly contradictory thrusts is the integrative task before us, selectively using conceptual roots from modern biology.

Biologist Richard Dawkins has stressed the role of cultural ideas which propagate themselves, termed "memes" after their analogy to genes. Similar conceptual leaps could dispel the physics-laden mechanical imagery which inhabits much of economics, politics and the humanities. Marxian ritual invocations of control and commodification, shrouded in a fog of conspiracy, proceed from misapplied classical mechanics. Self-organization through inherently chaotic interactions, the signature of our economic times, cannot emerge from the linear landscape of deterministic mechanical laws.

Replacing such habits of mind in the minds of managers requires such movements as "bionomics" and, generally, recalling that biology itself is no fixed set of immutable ideas. Evolution itself evolves as an idea, in our time introducing punctuated equilibria, sociobiology, and cladistics, the method of looking through time for linking evolutionary relationships among creatures.

All these will have to be tested, and in turn will generate fresh analogies for our views of economics. We shall see economic analogies which use not mere competition/cooperation balances, but complex, nonlinear responses to ever-changing circumstances. Predator/prey evolve into symbiotes; nature need not be red in tooth and claw.

Once our technologies learn the trick of reproduction a la nature, not a la factory, we may see a collision between the classical economy of scarcity and one of bio-plenty. Thinkers like Freeman Dyson have been pointing out that the specters which haunt our present — strip-mining and burning up our dwindling resources — may be as narrow a vision as was Spain's obsession with taking gold out of the New World, while missing tobacco, the potato, "love apples" (tomatoes) and the rest.

Biotech opens the promise that the truly fundamental resources will be sunlight, water, organic chemicals and land — privileging the tropical South and "green tech." This could neatly turn the tables on the industrial, "gray tech" North which will develop the biotech in the first place. (Spain sent Columbus, but missed the boat conceptually in the following century.)

An immense payoff for a small, but self-reproducing investment of "smart" biotech is a daunting possibility. We dropped jackrabbits into Australia before we knew their long-range impact. This point is not lost on the Luddites of our time, the Jeremy Rifkin crowd which fears any biotech product, and considers animal husbandry as "slavery."

The Wild Blue Maybe

One of the troubles with such apparently open-ended future projections is that we have no firm idea of what the limitations on biotech will be. Chances are, they'll be wilder than we think. The Frenchmen who first rode hot air balloons, gazing up at the lunar crescent, surely did not glimpse the century-long path that led through the airplane and the rocket to Tranquility Base.

The most complex riddle in biology is our own brains, possessing about a hundred thousand times the connections in a state-of-the-art Cray supercomputer. These connections work about a hundred thousand times faster than the comparable computer networks. This yields an organism with about ten billion times the capabilities of our billion-dollar number-crunchers. Consider what could be done by modifying some of the wiring diagram of that brain, or perhaps just some of its inherent chemistry. The potential for vast improvement or vast damage is immense.

Our currently common idea of software running on hardware works for machines, but not for brains. Our brains don't just store data in files. They modify themselves in response to strong inputs, laying down fresh patterns. This is why your memories of an incident can be modified by hearing another's version, or seeing a film of it. Brains form new routes for thinking—self-programming and self-hardwiring.

To reflect this, I think we will need a new category—liveware.

Like art, "living" is a property nobody can define exactly but everybody thinks they can recognize. The virtue of live technology is the same as the dray horse—it can look after itself, in its own fashion. Cropping grass, relieving itself, burning that grass for energy in its belly, the horse does a lot of its own maintenance. Liveware would similarly police up its own act, and be able to make copies of itself into the bargain, just like the dray.

A bioteched piece of liveware should be patentable. Europeans generally recoil from patenting any "living invention," whether a gene, a cell, an engineered plant or a human body part. In the U.S. patents are commonly given. We shall see a major collision between voices of environmentalism, "social justice" and religion—principally, the Pope—and international corporations like Monsanto. U.S. patent expert Rebecca Eisenberg recently observed that "In the U.S., we think of the morality issue as outside the realm of the patent system."

This fundamental division will accelerate as business tries to patent genes and plant traits extracted from the tropical world. The Department of Commerce estimates that life patents will be worth $60 billion worldwide by 2010. This is a deep, gut issue, like abortion, and it is approaching quickly.

But what's patentable is also, alas, mutable. Once made, it can undergo mutation and make something we did not intend. That's evolution, folks—biological, social, psychological. As Kevin Kelley's shotgun survey's title suggests, the essential quality of the Biological Century will be that in unsettling ways it will be Out of Control.

This very quality collides spectacularly with the pervasive scientific illiteracy of most industrial societies, especially the United States. Throughout the early 1990s the University of California, San Francisco waged a costly battle with nearby residents, simply to carry on their research—which they essentially lost. They were routinely accused of letting loose infectious pathogens, toxic wastes, and radioactivity. In public hearings, one excitable citizen suggested that doing recombinant DNA work had produced the AIDS virus. Another declared on television her outrage that "those people are bringing DNA into my neighborhood."

By the time the public begins to see just how rapidly change can come, the Easy Era will be over, never to return.

Human Categories

Yet intention is the crux of the moral issues we face. The abortion battles of our day will pale compared with the far more intimate and intricate capabilities that yawn just a decade or two away.

In the USA abortion won't go away as an issue, mostly because we keep trying to settle it through the courts. I suspect the Supreme Court will follow established practice and turn such a hot potato back to the states to decide, as they once tried to do with slavery. But that won't work when changes come thick and fast, as they are starting to.

Already Brahmins in India use amniocentesis to determine the sex of a fetus early on—and then preferentially abort the girls, because sons are more prized. This "genetic counseling" frames a typical conflict between our easy categories. Where does "reproductive choice" end when it systematically acts against females? If allowed to go on, we could produce harrowing population differences far from the near-50/50 balance of sexes, a testosterone-steeped society with more crime and war. I don't know the answers here, but I do know that the questions will get tougher.

And more subtle, as well. The first genetic tuning will be for the elimination of inheritable diseases—kidney disorders, hemophilia and the like. Such single-gene tailoring could appear around 2000. The Pope will oppose it as the opening wedge, and the battle will be joined.

It won't be settled, only broadened. For then will come genetic cosmetics: tailoring for eye and hair color, skin tint, maybe breast size (look at the implant industry today) and height. We do not know if these are controlled by single genes, but probably some are, and the others will prove to have only a few loci.

We're already familiar with the Yuppie competition to get Junior into the very best kindergarten. What expense will they spare if, a decade or two into the next century, parents can tailor their children for beauty? A firm jaw for men, a sunny smile for women? We all know that good-looking people do well. What parent could resist the argument that they were giving the child a powerful leg up (maybe literally) in the competitive world?

This will outrage many. Science is being perverted, they will say. From the noble elimination of a hideous disorder, like hemophilia, we will descend to the mere pursuit of transient appearances.

Somewhere, law (excruciatingly arrived at) or fashion (often underestimated, never absent) or deeper arguments (heard only by an elite) will draw the line. There will remain the familiar problem of oversubscription. Just as a Bachelor's degree was once a proud emblem, now tarnished by being commonplace, beauty—and, lest we grow smug, maybe even brains—will come to be so. Indeed, since beauty is another form of fashion, generations may sport characteristic, trendy noses and thighs, as now we see passing fads in children's names. But names can be changed.

Of course, the first genetic editing and rewriting will be done for the rich. When Scott Fitzgerald told Ernest Hemingway that the rich were different, Hemingway could confidently reply, "Sure, they have more money." No longer will that be strictly true. Rancor arising from built-in superiority, by right of inheritance, could soar.

One of our challenges will be to spread the benefit, or else see growing class separations of frightful complexity and depth. We could reach the stage in which one could spot the rich by their looks, or even their smarts.

Or their mates. Classical liberalism holds that information is good. And the truth shall make you free. If you can afford it!

Why, then, should a prospective bride not know the precise genetic endowment she would get from a candidate swain? We are just beginning to consider whether a genetic propensity for disease should be made known to insurance companies or employers.

Those legal battles can be settled in the context of privacy rights. But how about something as intensely personal as marriage? People care deeply about their children. It seems plausible that they would want to know what they are getting before going to the altar.

Being Human

All these naturally arising problems will tend to make us think of other people as anthologies of genetic traits—to atomize, a most thoroughly modern impulse.

This reflects science's tendency to slice and dice experience for convenience of analysis, but it is a poor model for knitting up the already raveled threads of a tattered society.

So somewhere, a line must be drawn. It had better be fixed by open public debate, rather than by our current method of leaving it up to lawyers in courtrooms, who usually know little and care less. Biology touches the wellsprings of our deepest emotions, making posturing before juries even worse than now.

Other developments, just over the horizon, will probably force us to entirely rethink present ideas of good and evil. Within a generation, we will probably be able to make cocaine from a bacterial culture. Kids will grow it or morphine or opium or marijuana—in bathtubs, not in elaborate labs.

This will do for our current drug prohibition what home-brewed beer did for Prohibition. Even easier ways are plausible: say, a bacterium which lives in your digestive tract and makes just the right level of cocaine every day. (Something like this has happened naturally. A patient turned up who was permanently drunk, from a yeast which made alcohol in his innards. Therapy freed him of a condition others might have envied.) Far more exotic methods of eluding detection, and of making new designer drugs, will no doubt emerge.

Such a ready supply will almost certainly doom a simple War On Drugs approach. Legalizing, taxing and regulating their use will come to be far cheaper than following a Prohibition mentality against an ever-improving biotechnology.

In fact, I believe it already is cheaper and smarter. We have over 1,300,000 in prisons in the USA, the majority for drug-related crime. The average sentence for murder in California is for fewer years (eight) than the average sentence for drug crimes.

Prohibition of anything is about power and imposing a uniform value system. Technology in the next century will probably act against central control. This will push our cultural boundaries, with biotech steadily worsening the friction. In the end this may force a new social solution, resembling the European programs already using partial legalization, combined with the social pressure which has reduced tobacco and alcohol use.

Dreams and Dreads

Out of playfulness—and cowardice—I've scrambled many ideas together without talking about when they might come.

To orient ourselves, I would call "mundane" the measures which have obvious market roles right away, and little social resistance. This includes pollution-policers, simple bathroom cleaners, crops that resist pests and herbicides, pharm animals, "designer" plants (blue roses, low-cal fruit), bacterial mining, and the like. Even correcting human inheritable diseases will probably go through without major opposition. All this, perhaps within the first two decades of the new century.

The battles will begin in earnest with conceivable but startling capabilities. The list is long. Big changes in our own genome. Harnessing natural behaviors to new tasks (the acacia ant–orange tree marriage). Designer animals, like a green Siamese cat to match your furniture, or even a talking collie (and what would it say?). These may preoccupy the middle of the next century.

Even further out would be major alterations in the biosphere, and in us. Adapting ourselves to live in vacuum or beneath the sea, or to convert sunlight directly into energy, would alter the human prospect beyond recognition. Changing homo sapiens to something beyond will be a step fraught with emotion and peril. Such issues will loom large as the Biological Century runs out. And what could lurk beyond that horizon? The mind boggles.

All these are mere glimpses of what awaits us. A century is an enormous span, stretching our foresight to the full. Reflect that H.G. Wells' *The Time Machine* appeared only a century ago, in 1895. Biotech can usher in as profound a revolution as industrialization did in the early nineteenth century. It will parallel vast other themes — the expansion of artificial intelligence, the opening of the inner solar system to economic use, and much, much more.

The Achilles heel of predictions is that we cannot know the limitations of a technology until we get there. A nineteenth century dreamer might easily generalize from the forthcoming radio to envision sending not merely messages by the new "wireless," but cargoes and even people. Matter, after all, is at bottom a "message."

But there's more to it than that: the awesome radio didn't develop into a matter transmitter, which is no closer to reality than it was when it was first suggested.

So undoubtedly I'm wrong about some of these analogy-dreams, particularly the timing. What I will bet on is that, despite the current fashion for "nanotechnology" — artifice on the scale of a nanometer, the molecular level — biotech will come first. It is easier to implement, because the tiny "programs" built into life forms have been written for us by Nature, and tested in her lab.

In fact, some of the most interesting prospects of nanotech-like thinking come from biological materials. The basic mystery in biology is how proteins figure out how to fold themselves, which determines myriad biological functions. An obvious long chain molecule to fold and use as a construction material is DNA itself. A self-replicating "bio-brick" could be as strong as any plastic.

Consider adding bells and whistles at the molecular level, through processes of DNA alteration. Presumably one could then make intricately malleable substances, capable of withstanding a lot of wear and able to grow more of itself when needed.

It isn't fundamentally crazy to think of side-stepping the entire manufacturing process for even bulky, ordinary objects, like houses. We have always grown trees, cut them into pieces, and then put the boards back together to make our homes. Maybe we will someday grow rooms intact, right from the root, customized down to the doorsills and window sizes. Choose your rooms, plant carefully, add water and step back. Cut out the middleman.

Whether such dreams ever happen, it seems clear that using biology's instructions will change the terms of social debate before nanotech gets off the ground.

Hello, Dolly

I am a clone.

Or rather, I am better than one. Or so any identical twin surely must see the matter.

The recent media feeding frenzy about a cloned sheep, Dolly, showed us journalism in its fullest modern form. I conclude with this example, for it tells us much about how we scientists inevitably will be portrayed.

Many of those writing about this genuine watershed moment in techno-culture followed current journalist practice: their foremost research instrument was the telephone. Of those who called me — an unlikely authority, since I am not a biologist — none realized that DNA does not solely determine the heritage a child gets from its parents.

My brother, Jim, and I shared a womb without a view for nine months. (Though not always restfully, our mother reports.) Genetically identical, we also enjoyed the same currents and chemicals of our mother. After a rather traumatic birth — both had our appendixes removed within days — we were brought up in the same house, with constant attentive parents, and even wore matching clothes until our late teens. (How much trauma this clothing ritual induced in our personalities I leave to others to decide; suffice to say that being seen as sugary-cute has left me with a decided prejudice against sweets in any form.)

True twins share womb chemistry and endure many fateful slings and arrows together. The fabled connection between twins is true, in my case. We are distantly dismissive of mere fraternal twins (different DNA) and regard all others as "singletons," those condemned by birth to endure the isolation of never truly sharing the intuitive grasp that we enjoy without paying a price.

Or nearly so. There is mild statistical evidence that identicals have slightly lower IQ. This might be plausibly so; the comfort of ready communication may well lead to a certain mental laziness.

Jim and I felt the opposite. Reared in rural southern Alabama, we enjoyed an idyllic Huck Finn boyhood. But education there was casual at best. Our mother and father were high school teachers, and challenged the pervasive easy-going ignorance. We attended a one-room schoolhouse, with each row of seats a separate grade. Against this my brother and I united, reading widely and enjoying the clash of cultures which paraded by. After we were 9 our father became a career Army officer, whisking us to Japan for three years, Germany for another three, and further isolating the twins from a continuity that might have sucked us into the conventional.

So we are an odd pair even among twins. Jim got his doctorate from the same institution as I, UC San Diego, in the same area (plasma physics) and now lives a few kilometers from where I once lived, in northern California. Such correlations appear often among twins. We grow up in a culture of sameness, so have a sense of self always shared.

Among singletons, interest in twins is enduring. Do we feel some mystical sense of connection? Of course; but whether it is mystical or not begs description. I am writing this at 35,000 feet over Greenland, on the way back to UC Irvine from Lapland. I know without thinking about it that my brother is probably body surfing on a beach near La Jolla, though I have not spoken to him for ten days. I remember his itinerary and without conscious deliberation feel where he is likely to be. This is processing at the unconscious level, and as an experiment, when I see him in two days I shall check with him and let you know the outcome. [Later: my estimate was right to within the hour.]

But this is scarcely mystical. Instead, I attribute the innumerable similar incidents in our lives to a lot of automatic thinking, based on intuitions cooked up through more than five decades. To singletons this can look uncanny.

Speaking as a twin, clones seem a lesser form. They grow up in a later era than their genetic duplicates, with different upbringings. Would knowing that they were genetic duplicates trouble them? Surely such people would not be inherently more mentally fragile; Siamese twins are far more like each other than ordinary twins, yet suffer no higher incidence of mental illness than is usual, suggesting that even extreme parallels in nature and nurturer are not damaging.

The furor over Dolly puzzled me by the emotional level of debate. Reasonable people like political commentator George Will asked, "What if the great given — a human being is a product of the union of a man and a woman — is no longer a given?" This issue properly comes from a broader issue in biotechnology, the entire field of artificial birth in all forms, for there are no precise boundaries in this new territory.

Certainly I see no reason why society should prevent grieving parents from having a baby cloned from the cells of a dead child, if they wish. Beyond such emotionally wrenching cases, where should we erect walls? Oxford biologist Richard Dawkins asserted that he could see purely intellectual issues intriguing enough to justify cloning himself: "I think it would be mind-bogglingly fascinating to watch a younger edition of myself growing up in the twenty-first century instead of the 1940s."

Many no doubt find his position puzzling or even immoral or disgusting. Even so, why should Dawkins be prevented from having a cloned child? What is society's mandate?

The Dolly debate produced several claims that cloning violated the fundamental principle of individual dignity. Twins certainly belie that argument. Fears of interchangeable people armies, usually marching robotically onward, come from a simple-minded genetic determinism. And the grounds for a principle of uniqueness seem vague at best.

After all, why treat clones differently? — we twins and clones are all "monozygotes," as the biologists put it. In fact, clones necessarily separate in outside influences from their first moments in the womb, for the wombs are different. Another's DNA inserted into a host egg will acquire "maternal factors" from the proteins of that egg, affecting later development. The womb's complex chemical mix varies with each mother, so nine months of different "weather" will change the outcome in the fetus; the baby will not be a photocopy of its older original.

And clones will be full-fledged people with all rights attendant to that status. Nobody forces twins to serve as organ farms for their other twin; clones would have the same legal status.

The true first use of cloning will undoubtedly be in the "copying" of highly selected farm animals. These could first be excellent milk cows or racing horses. More futuristically, we shall see — and quite soon — the cloning of "pharm" animals which yield biotech products of use to us, such as insulin-rich milk from cows, and a whole array of therapeutic hormones, enzymes and proteins.

Plants have already been extensively engineered. More than three-quarters of the cotton grown in Alabama last year was genetically tuned to kill predatory insects. Already scientists are experimenting with cotton plants that contain polyester fibers, too, surely a boon for fans of leisure suits.

Still, cloning should indeed furrow the brow of long-perspective thinkers. We believe sexual reproduction holds sway over much of the kingdom of life because it provides ever-new gene mixes, allowing a species to build fresh defenses against the ever-mutating pathogens that infest the natural world. The perpetual arms race between prey and predator favors sex as a defense. Seen this way, we are men and women because the primary predators on humans have always been microbes, not tigers.

So "pharm" animals cloned repeatedly will face the very real threat of infectious diseases which wipe out a herd overnight. But surely nobody will clone huge numbers of humans, so such plagues will be unlikely. The breeds of influenza that regularly attack us genetically diverse humans will do far more damage.

The Ethics of Bioethics

As I write this, a presidential panel seems about to recommend a uniform federal ban on human cloning experiments. I believe this will be a mistake, generally, and an ineffective move anyway. The technology is fairly simple; others will pick it up. In Latin American countries or on offshore islands, clinics will offer the service at a hefty charge. Underground, without legal oversight, we will indeed see some tragedies and even horrors.

Bioethics is a field with many practitioners but few obviously qualified savants. Often the bans which spring from such federal committees prove ill-advised, their only long-term effects negative. This was the case with the two-year moratorium on recombinant DNA, which simply slowed the field without deciding anything. So did similar bans on selling organs or blood, and I predict, so shall the recent Clinton prohibition on using human embryos in federally backed medical research. The ultimate price for these momentary interruptions—and so far they have always been momentary—is lives lost because the resultant technology arrives too late for some patients.

Bioethicists tend to see problems everywhere, and saying "no" gives them visible power. Letting technology evolve willy-nilly, responding to what people want—maybe even people without advanced degrees!—gives bioethicists no perks or prominence; unsurprising, then, that they seldom go that route. They aren't the patients clinging to life, or infertile, or stunted in some potentially fixable way.

They also tend to think collectively, omitting the inconvenient needs of real people. Bioethics professor George Annas of Boston University flatly demands, "I want to put the burden of proof on scientists to show us why society needs this before society permits them to go ahead and [do] it." Note that he does not require this rule in his own work, including testing the above sentence by its own standards. Instead, Virginia Postrel has noted of Annas and many others, "We will hear the natural equated with the good, and fatalism lauded as maturity. That is a sentiment about which both green romantics and pious conservatives agree."

Indeed. We would save ourselves much trouble if we could agree that the proper place for most bioethical thought lies in counseling those affected, not in dictating the spectrum of possibilities.

In Conclusion

We view ourselves through the lens of morality and ethics, too. Surely the bedrock ideas evolved by hunter-gatherers and simple farmers are not going to work much longer, in the face of such fundamental change. At the far end of the next century looms a horizon that is for me quite out of sight.

The rate of change of our own conception of ourselves will probably speed up from its present already breakneck pace. The truly revolutionary force in modern times has been science, far more than "radical" politics or the like. This seems likely to be even more true in the future.

Yet the above examples underline the implications of leaving genetic choices to individuals. Society has some voice in defining boundaries, surely. But typically we arrive at consensus slowly, while biotech speeds ahead. Perhaps here we see the beginnings of a profound alteration in the essential doctrine of modern liberal democratic ideology. There may be genetic paths we will choose to block. How do we recognize them, quickly?

Our species has made enormous progress through swift cultural evolution. Now that quick uptake on changing conditions can come from deep, genetic change. We will hold the steering wheel, however shaky our grip, and not give way to pitiless, random mutation.

We will emerge from a Biological Century with a profoundly different world view. Perhaps some new technology will promise to shake our foundations in the dawn of the 2100s, too. Mercifully, we cannot see that far.

Our prospect is wondrous and troubling enough. It is as though prodigious, bountiful Nature for billions of years has tossed off variations on its themes like a careless, gushing Picasso. Now Nature finds that one of its casual creations has come back with a piercing, searching vision, and has its own pictures to paint.

4

THE CAMBRIDGE QUINTET: THE CHRONICLE OF AN EXPERIMENT IN 'SCIENTIFIC FICTION'

John L. Casti

The Birth of a Book

Sometime back, I attended a scientific meeting in New York City at which the scientists present tried to convince their peers of the virtues of their particular theories by creating what can only be described as stories. Of course, the tellers of these tales didn't see them as stories at all, but only as versions of "The Truth" in which various inconveniences of the real world had been simplified—or simply omitted—so as to make the story come out right in the end. Such idealizations of nature are a common enough practice in science, although most practicing scientists would probably balk at the labeling of their theories as "stories." But, in fact, in their distortion of real-world facts these stories are every bit as fanciful as the types of warpages of reality engaged in by novelists in creating their tales of human strife and struggle.

Having had enough of these scientific stories to last me for awhile, one afternoon I skipped the meeting to engage in my absolutely favorite pastime in New York—haunting the bookshops from uptown to downtown and most everywhere in between.

During this welcome break from scientific storytelling, at one shop I ran across the published scripts from one of the most successful television series ever produced on PBS, the series *Meeting of Minds* [1]. This Peabody Award–winning series hosted by the polymath, Steve Allen, ran from 1977 to 1979. The format of the show was an interesting one. Allen played the role of the host of a dinner party, to which he would invite famous characters from history—Emily Dickinson, Attila the Hun, Charles Darwin and Vladimir Lenin, for example. The content of the show was then simply a dinner-party conversation, ranging over a host of topics that Allen, like any good host, would periodically inject into the discussion.

As I thumbed through these scripts, recalling the pleasures I had when I'd watched the shows in their original screening, I wondered about whether it might not be possible to present ideas of both scientific and philosophical content in this dinner-party format. Why couldn't one, for instance, have a fictional dinner party to which historical characters would be invited to discuss questions like the origin of life, the possibility of intelligent extraterrestrial life, or the nature of quantum reality? After purchasing the four volumes of TV scripts, I continued pondering this question as I read them more carefully on the long flight back home. Surely, there must be a way of sugar-coating the scientific message of some of the almost impenetrable talks I'd heard earlier in the week in some sort of fictional format, so that the basic idea could be presented in a comprehensible and entertaining way for an intelligent non-specialist. Thus was born the idea underlying my 1998 book, *The Cambridge Quintet* [2]. More on that in a moment.

Science as Storytelling

Both the creation of science and its recounting for non-scientists almost always takes place in the form of stories. For example, scientific theories of the origin of life are developed around "scenarios" by which molecules floating around in the "primordial soup" self-assemble somehow into self-replicating entities from which the myriad lifeforms we see on Earth today have evolved. Similarly, one of the great mysteries in quantum theory, how the Schrödinger wave function, a purely mathematical entity, describes the measured behavior of a physical object like the momentum of an electron, is puzzled out through Schrödinger's story of a cat in box, the so-called Schrödinger's Cat thought experiment. Both the origin-of-life scenario and Schrödinger's Cat are stories that scientists tell to each other to clarify and extend theories of the physical universe. But there are also the stories scientists tell to the lay public about their work.

The stories of science that a layperson encounters may take many forms. Let me take a longer look at just three of the most common types.

• *Expository Articles:* The Op-Ed page and Science sections of major newspapers, such as *The New York Times* or *The Washington Post*, often feature articles written by scientists, whose aim is to tell a story illustrating one or another development in contemporary science. For instance, a few months ago I wrote such a piece myself, in which the aim was to give an account of how it could be that well-meaning—but hardly uninvolved—climate researchers could come to such diametrically opposed conclusions on the matter of global warming. The story I told in this instance was a common one for scientists, involving the problem of bringing computer *models* of climatological phenomena into congruence with their real-world counterparts.

Scientific exposition can also be directed at other scientists, as for example in the survey articles and more technical expositions found in general-science periodicals, such as *Nature, New Scientist, Science,* and even *Scientific American.* Again, such stories aim to inform their target audience by creating stories that have enough detail to be informative, but not enough to become tedious for non-specialists.

• *Popular Books:* Nowadays, just about every general bookstore contains a separate section labeled "Science." And the overwhelming majority of books on display in this section are written by scientists (although not always *practicing* scientists). Occasionally, such a book even achieves a kind of cult status, as with the volumes *Gödel, Escher, Bach* [3] by Douglas Hofstadter and *A Brief History of Time* [4] by Stephen Hawking. In both these cases, the volume told a story about some aspects of the natural and human worlds, that was designed to convey information to a non-specialist reader about matters scientific.

• *Fiction:* Certainly the least explored—but potentially most valuable—mode of scientific storytelling is through the medium of fiction. And I don't necessarily mean "science fiction," but rather general literature that takes science and/or scientists as its focus. John Updike's volume, *Roger's Version* [5] is a notable example in this regard, along with the more recent volume *Enduring Love* [6] by the British novelist Ian McEwen. These works happen to be outstanding pieces of literature that just happen to be about science or scientists, in much the same way that the novels of John LeCarré happen to be about the British intelligence community. Occasionally, such volumes are even produced by scientists themselves, as with *Cantor's Dilemma* [7], the first volume of a tetralogy by noted biochemist Carl Djerassi, that he terms "science-in-fiction." And, of course, there is the more mainstream science-fiction literature, about which so much has been said that I will not discuss it further here, other than to note that many of the classic volumes are difficult to distinguish from the type of science-in-fiction that Djerassi has written. If there is a distinction to be made, my observation is that it resides in the relative emphasis placed on the motivations and character of the scientists in Djerassi's works as opposed to the exploration of the implications of the science, constituting the principal focus of more traditional "hard" science fiction.

While I find all of these modes of telling stories in the cause of science fascinating and of great value—both as enlightenment as well as entertainment—it seemed to me that there was yet another possibility, one that I like to term "scientific fiction." Let me outline the basis for this form of science-as-storytelling, at least in the form I thought of it when producing my book, *The Cambridge Quintet.*

Scientific Fiction

The greatest single attraction of an historical novel is that it deals with *real* historical characters in a "what-if?"-type of setting. So readers may not only be entertained, they may also actually learn something about how these people of the past lived, worked and played. More than anything else, it is this emphasis on real personages interacting over real intellectual issues that distinguishes what I call "scientific fiction" from the modes of presenting science outlined above.

A work of scientific fiction is composed of four principal parts.

- *Historical characters:* The *sine qua non* of scientific fiction is a collection of historical characters whose lives, thoughts, and interactions carry the story of the book. Since the focus is on matters scientific and philosophical, in such a work these characters will mostly be drawn from the worlds of science and closely related intellectual areas. However, this constraint certainly does not rule out artists, writers, musicians, or other types of thinkers besides scientists playing an important — but secondary — role in the story.

Above all else, the focus of a book of scientific fiction is on explicating scientific issues, not on the characters. It is for this reason that a work of scientific fiction is *not* a novel, at least not as that term is generally understood. But it is a work of fiction. Some readers — and reviewers — of my book [2] seem somehow to have suffered some confusion on this point. A work of scientific fiction should be judged on how well it presents and explains the science, not on its development of the characters. The interaction of the characters is important, of course, but only insofar as it pushes along the goal of illuminating the intellectual issues residing at the heart of the book. Beyond that it's only an unnecessary and unwanted distraction.

- *Place(s) and Time(s):* The historical characters need a playing field in space and time upon which to act out their struggle with the scientific themes of the story. The only real constraints in this regard are that the place and time be consistent with the issues being considered. For example, it would not do at all to set a story about genetic engineering at Plato's Academy in ancient Athens. Nor could an account of the phlogiston theory of matter be credibly placed at the Cavendish Laboratory in 20th-century Cambridge.

- *Surface Theme(s):* As with any good story, a scientific fiction requires a point of conflict over which the characters can interact. I've found it useful to have at least two sorts of conflicts, a sort of surface conflict that may or may not be of ephemeral importance and a deeper conflict that generally rests upon one or another of the eternal conundrums of philosophy. The surface theme(s) might revolve about matters as trivial as academic bickering over rank and position or something a bit deeper, but equally specific, such as a basic mystery of science like the origin of life. But the main point is that the surface theme be sufficiently contentious — and urgent — that it can both capture the reader's attention and serve to keep the story moving along at a brisk pace. In summary, the surface themes emphasize the emotional and "human" issues of the story.

  • *"Deep" Theme(s):* The deep theme(s) of a scientific fiction relate to the surface theme by serving as a general foundation upon which the surface theme rests. In other words, they create the philosophical and intellectual underpinnings of the story. For instance, if the surface theme were the question of how life originated here on Earth, a possible deep theme supporting a debate on various theories of the origin of life might be the uniqueness of human physico-chemical structure. In other words, would we expect to find the same sort of carbon-based physical structures wherever life has originated in the universe or is there something special or accidental about how it happened here on Earth? Whatever the deep theme of the book happens to be, it should be the main intellectual underpinning of the entire enterprise and serve to support not only the surface theme, but also the interaction of the characters as they debate and discuss the pros and cons of various possible resolutions of the surface theme(s).

This discussion of people, places and themes has been quite general. So let me bring it down to earth by showing how the above structure is implemented in the specific case of my 1998 book, *The Cambridge Quintet.* The pretense of this work is a fictional dinner-party conversation at Christ's College, Cambridge, in June of 1949. The host of the party is the novelist, scientist and government mandarin, C. P. Snow, who has gathered a star-studded collection of thinkers to consider the problem of whether a computing machine can be constructed and programmed so as to duplicate human cognitive processes. To debate this matter, Snow has invited the philosopher Ludwig Wittgenstein, the mathematician Alan Turing, the physicist Erwin Schrödinger, and the geneticist J. B. S. Haldane. The matchups between the general categories for a scientific fiction outlined above and their implementation in *The Cambridge Quintet* are given in the box below:

Category	The Cambridge Quintet
Characters	Alan Turing, C. P. Snow, J. B. S. Haldane, Ludwig Wittgenstein, Erwin Schrödinger
Time and Place	Christ's College, Cambridge, June 1949
Surface Theme	Can human cognitive processes be duplicated by a computing machine?
Deep Theme	The uniqueness of human cognition

Several people have asked me how I came to choose these particular characters and this time and place as the basis for the book. The answer is that once I had settled on the theme of the book (thinking machines) and the time (mid 1949), the rest was almost foreordained. It was clear that work on computing machinery and code-breaking during World War II had set the stage for the thinking-machine question, which was finally elucidated in Alan Turing's famous paper "Computing Machinery and Intelligence" [8], which was published in 1950. So it's reasonable to suppose that Turing was putting the finishing touches on this work in 1949; hence, the timing of the dinner party and one obvious guest. As Turing's foil it was natural to select Wittgenstein, who not only held a view of human intelligence that was anathema to a mechanist like Turing, but was also Turing's teacher in the sense that Turing was Wittgenstein's target audience in his lectures on the foundations of mathematics held in Cambridge in 1938–39 when Turing was a student at King's College.

As for the other guests, I wanted a representative from both the life sciences and the physical sciences in order to shed light from these areas on the philosophical and psycho-sociological aspects of artificial intelligence. As perhaps the most philosophically-minded of all the great quantum physicists, Schrödinger was a natural choice to represent the view of the physical sciences. Similarly, the geneticist J. B. S. Haldane was a logical choice to speak for the life sciences, especially in view of his far-ranging interest in the social aspects of biology.

Finally, the host. For this role I needed someone who understood science but was really a humanist. The novelist C. P. Snow fit this bill perfectly. He was not only a trained chemist, but served for many years as a government advisor on matters of science and science policy. And since Snow was the host of the meal, what better place to hold it than at his old rooms at Christ's College, Cambridge? This venue seemed doubly appropriate, since both Haldane and Turing had been students at Cambridge, while Wittgenstein had been a professor there.

So that's how the skeletal structure of this particular work of scientific fiction came about. Just to show that this lineup of people, places and ideas is not special, let me close by outlining the structure of another book of this type, one I have tentatively titled *The One, True, Platonic Heaven*. This volume is set in Princeton at the Institute for Advanced Study (IAS) just after the Second World War. The principal characters are the mathematician John von Neumann, the logician Kurt Gödel, the physicist Albert Einstein, and the director of the IAS, J. Robert Oppenheimer. There are two surface issues serving to carry the story in this work, the first being the enormous resistance of the IAS faculty to von Neumann's proposal to build a computer at the Institute. The second surface theme, which is strangely intertwined with the first, is the faculty's puzzlingly long debate over whether or not to promote Gödel to the rank of Professor. The oddness of this situation is that Gödel had already been universally recognized as the greatest logician since Aristotle. So why did the faculty wait so long to promote him to Professor?

Underlying these surface themes is the general philosophical question of the limits to scientific knowledge. Gödel had opened this book with his famous result on the incompleteness of mathematics, showing that even in mathematics there are no mechanical "methods" for discovering all true statements about numbers. This fact is intimately tied up with the notion of a computation. So if one wants to know about limits to scientific reasoning *outside* mathematics, then it's necessary to confront the question of limits to computation. At this point the story intersects von Neumann's quest to build the world's first "supercomputer." The following table encapsulates the main ingredients in this story:

Category	*The One, True, Platonic Heaven*
Characters	John von Neumann, Kurt Gödel, Albert Einstein, J. Robert Oppenheimer, Hermann Weyl
Time and Place	Institute for Advanced Study, Princeton, 1946
Surface Themes	von Neumann's computer project; Gödel's promotion to Professor
Deep Theme	The ultimate limits to scientific knowledge

Besides the formal structure of a time and place, real-life characters, and surface and deep themes, what *The Cambridge Quintet* and *The One, True, Platonic Heaven* have in common is the attempt to elucidate and explain philosophical ideas of substance in a semi-fictional format. In this sense, these books are much more like Jostein Gaarder's "children's story" *Sophie's World*, [9] than they are like a mainline novel. Despite the fact that the subtitle of the book calls it "A novel about the history of philosophy," the book is a philosophy course set in one typeface interspersed with a fantasy printed in another. But the fantasy is only the sugar-coating to make the philosophy palatable. So it is too with the books of science I've outlined here. To paraphrase Marshall McLuhan, the ideas are the message. Hopefully, the success of Gaarder's work, as well as the hoped-for readership of *The Cambridge Quintet*, will encourage authors more skillful than myself to join in the fun. The world of science and philosophy has lots of stories to tell, and "scientific fiction" is a new way to tell them.

References

[1] Allen, Steve. *Meeting of Minds.* Series 1–4. Buffalo, NY: Prometheus, 1989.

[2] Casti, J. *The Cambridge Quintet.* London: Little, Brown & Co., 1998 (American edition: Reading, MA: Perseus Books, 1998).

[3] Hofstadter, D. *Gödel, Escher, Bach.* New York: Basic Books, 1979.

[4] Hawking, S. *A Brief History of Time.* New York: Bantam, 1988.

[5] Updike, J. *Roger's Version.* New York: Knopf, 1986.

[6] McEwen, I. *Enduring Love.* New York: Doubleday, 1997.

[7] Djerassi, C. *Cantor's Dilemma*. London: Macdonald, 1989.

[8] Turing, A. "Computing Machinery and Intelligence." *Mind*, 59 (1950), 433–460.

[9] Gaarder, J. *Sophie's World*. New York: Farrar, Straus and Giroux, 1994.

5

BECOMING MAUREEN— A STORY OF DEVELOPMENT

Jack Cohen

There is a great part of all languages devoted to processes, additional to those usages for things and for events, and we find those concepts, even the words, difficult. Nouns like *embryo, child, adolescent*, abstract nouns like *development, construction, flight*, verbs like *to become, to grow*, even *to die* are very difficult to handle mentally for most people. We cannot imagine becoming nothing when we die, and the concept of a continuing soul is a way out of that inability; similarly, early microscopists "saw" a homunculus in the spermatozoon, which saved them from having to imagine a real increase of complexity during development.

The major problem most lay people have with the evolution of animal and plant adaptations is "But what did it have to start with?", sometimes forced into the question "What use is half an eye?" Dawkins [1] has beautifully illuminated that one with "Any light-sensitivity is better than none!" and Nilsson and Pelger [2] have shown how easy complex eyes are to develop over mere thousands of generations. Nature, too, finds it easy: many different evolutionary lines have done it. Dawkins' explanation of the gradual acquisition of camouflage to answer the question "How did an insect come to look so perfectly like a twig, or a leaf?" starts with a poor-sighted, not very hungry predator in twilight! Even Stephen Jay Gould, evolutionary biologist par excellence, quoted the flamingo's feeding with its head upside-down as a sudden acquisition, to be explained [3]; but I can easily imagine an ancestor which sometimes turned its head upside-down, or even sideways, to get insect larvae in shallow water ...

We talk of characters in novels developing, we even talk of story-lines developing. Adrian Mole [4] does so convincingly, in contrast perhaps to Pooh Bear [5]. We speak of sciences developing, classically [6] as Newton is replaced by Einstein (note the "replacement," an easy idea, supplanting the gradual process). But it is in the biological sphere that processes are of the essence. Having taught University students, and lectured to many lay audiences over the years, I have met the difficulties. They were made very manifest in the Eldredge and Gould suggestion of punctuated evolution: [7] they meant short-term gradualism between long periods of several kinds of morphological equilibria, but the media—and some other kinds of scientist less familiar with palaeontology—interpreted this as stasis and replacement. Like Einstein replacing Newton [6], that is easier to think about.

People, and I include most embryologists, find it hard to come to terms with the real increase of complexity during development. For example, however we measure complexity, a growing feather is more "complicated," more complex, than the patch of cells on the yolk from which the whole chicken arose. We, and the chicken, genuinely do complicate as we develop—and we invent concepts to explain this away [8]. Preformationisms are the most usual: "it has always been that complicated, only you couldn't see it!" If there's a little man in the sperm, it only has to get bigger; if there's an entelechy (Driesch [8]), it has only to find extension in space; if there's a DNA blueprint it just has to be actualized in flesh. "Conservation-of-information" seems to be our default. We are much happier believing in transformations, in information which is behaving like proper Shannon-Weaver information should [10]: being transduced into the transmitter by a code, received and decoded at the receiver. We even speak of the DNA message "passing across the generations" when there is no receiver; indeed, the implied function of the message is to build the receiver. The information-metaphor falls apart in all kinds of ways; for example, there is not enough information in a mammal's genome to specify the neural connections of quite a modest mammal, never mind the human brain. But it does make a good story; it sounds as if it works. So did the little man in the spermatozoon, so does DNA-made-flesh in the modern equivalent. In this view, Maureen is the realization of her DNA blueprint.

We need new metaphors for development, and there have been many attempts to invent them, to progress from the conservation-of-complexity models into the real world, where complexity is lost or gained all over the place. I am intrigued by a very obvious real-world example, the way flames re-invent the morphology in each "generation" (as each new combustible is lit), accessing the same set of rules each time. To some extent, of course, this is what chick embryos do, and the complexity of feathers reflects constraints on the geometries of feather development.

Stuart Kauffman's N/K networks [11], and many other models of genes-controlling-development-with-real-world-complexity persuade us that organization, complication and life are downhill from chemistry. This is consonant with much kitchen-sink chemistry (caramel is much more complicated chemically than the sugars and fats which were cooked to make it) but not at all congruent with what we remember of school chemistry. Chemistry lessons in school have to be "terminal" or "equilibrium," to fit the practical classes. But real chemistry, of clays, rusting, ozone layers and smogs is very recursive, with many autocatalytic steps — even a hydrogen-and-oxygen mixture isn't explosive when it's dry; it requires some water to be autocatalytic, and then many chemical steps to finally explode into water. So the old problem of the origin of life, when we found it necessary to build complicating chemistry in great arrays above the "simple" prebiotic chemistry, until we reached the complicated height of the simplest biochemistry and it could all get started, was another misplaced problem of process [16]. Much chemistry is recursive, and it could become life by linking up its circles to include local substrates and energy sources; biochemistry didn't have to replace chemistry (like Einstein/Newton), the chemical processes became (another one of those difficult concepts) living.

Homeostasis is a very common explanation of self-maintenance of physiology: organisms have negative-feedback systems which maintain concentrations of ions, numbers of enzyme molecules, areas of membranes and for "higher" organisms, maintenance of temperature. Homeostasis has, as metaphor, a ball in a cup. We understand much of physiology and biochemistry as systems which are self-maintaining; a thermostat is a useful metaphor—the ball rolls into the bottom of the cup again if it is displaced. Bacteria do not develop, they continue to do the same processes (replicating DNA, synthesis of membranes and ribosomes) which result in multiplication of the same entities. In contrast, embryology does not result in maintainance but in progression: the egg becomes an embryo, the embryo becomes a fetus, a baby, a child, an adolescent who becomes an adult. This development is a special evolutionary invention, originated on many different occasions and by different mechanisms. Bacteria make more bacteria by doing homeostatic things, by and large. But Bactrian camels don't make more Bactrian camels (although they do have lots of fabulous homeostasis ...), they make eggs and spermatozoa whose combination fertilizes, which proceeds into cleavage which makes a blastocyst, which is the kind of physico-biological system which makes an embryo which burrows into mother's uterine wall, which then makes blood vessels so that it may better parasitize the mother; it then becomes a fetus, which grows until it stimulates the mother to initiate the birth process. The baby camel is the kind of system which forces its mother to feed it, so that it grows into a juvenile, then an adult. Only then can the resultant successful adult make eggs or spermatozoa again. Reproduction of bacteria is homeostasis in many dimensions. But reproduction of birds or oak trees or mushrooms has to proceed, develop, through many kinds of systems to come around to the original stage again. This is not ball-in-a-cup, homeostasis, but ball-rolling-down-a-valley, homeorhesis [12].

Waddington [12] made a heroic attempt to describe development in different terms from physiology. He did many experiments showing how modifiable development was, in ways which disconfirmed the "Have I got the right mutation?" image of naive genetics. "Chreodes" were an attempt to see embryology as a series of trajectories of different parts of the embryo as they diverge, frequently interacting as they mutually proceed by homeorhesis. The parts of an embryo run down the bottoms of their notional valleys (like rivers); and if, by chance or flood, the course of development runs up the side of a hill it never goes back to what it was, it runs down to the valley further down toward the sea. He complicated this idea into the "epigenetic landscape," a developmental metaphor whose function was to explain animal development by homeorhesis, not homeostasis: balls rolled down valleys in a landscape whose topography was created by genetic interactions from underneath (the genes shown as webs pulling each others' strings) while the environment pressed the landscape from above.

There have been many developmental metaphors which have grown from this delightful Waddington picture. Some of the most useful emphasize that the landscape changes as the organism develops, new rules appear because development has proceeded into new circumstances. Goodwin [13] has several models of this general kind. He has attempted to fit developmental paths into the phase space of possible sequels to eggs, constraining them by geometry on the one hand and the smallish repertory of proteins from DNA on the other. The successive stages exist in new sets of rules, so that "the same" DNA-making-protein can have quite different effects later, in a wing bud for example, in contrast to segment formation in the early embryo. The phase space of possible effects of transcription of any gene is dependent upon which place, and at which stage, it happens. In this kind of vision of development there is therefore [14], in principle, no mapping from a particular gene to some structure in the final stages of the organism. This kind of metaphor for development contrasts with that of Dawkins [15], whose "gene for a nose," or for a wing or for a hand, draws the reader to suppose that the DNA genome is, in some sense, the pattern for the later animal. Such a "blueprint" model, like all pre-formationist explanations of development, is intellectually lazy. Some few embryologies (*Caenorhabditis* and other nematodes, *Arabinopsis* and a few other plants) do seem to have nearly blueprint development, but nearly all complex organisms from oaks and dandelions to zebra-fish and zebras, have interactive homeorhetic development whose trajectory changes the developmental rules as it proceeds. This kind of complicity [14] between the genome and its environment, including its read-out mechanisms, precludes mapping from genome to character. (But, of course, one can map *differences* of genetics to *differences* of character—all else being equal; that is what geneticists used to do, in *ceteris paribus* mode, and why it is reasonable

to speak of genes for albinism in mice or for vestigial-wing in
Drosophila.)

Ian Stewart and I have been very concerned to explain this
real problem so that people can genuinely understand how it
comes about that a simple egg can routinely make a complex
organism — how one feather can be more complicated than the
germ of the whole bird — or how it comes about that Maureen can
grow up. We played with phase spaces [14], and the emergence of
complex properties, particularly as developmental phase spaces
modify themselves into new rules like Langton's Ant (simplexi-
ties), or phase spaces of different systems interact and make new
possibilities altogether (complicities). We also used game trees —
in which any embryo plays one of the possible chess games, and
even the whole species only actualizes a very small volume of the
possible [16]. Metaphors can be much simpler, but still useful. We
like car routes on landscapes, which demonstrate homeorhesis:
"We went wrong three junctions ago, but we can take a left at the
next lights, turn right at the third crossroads and that will bring
us back to where we would have been if ... "; a whole journey,
with its possible by-roads and corrections, is like a development.

In our view, which contrasts with preformation of any (including DNA) kind, Maureen starts as an embryo, but baby Maureen begins to control her environment, to force her mother's
smile and her mother's milk. Maureen lives through, and is
developed by, the succession of environments which in our culture makes up the Make-A-Human-Being Kit [16]: nursery songs
and stories with iconically sly foxes (in contrast to Inuit foxes,
brave and fast; or Norwegian foxes, subtle and full of good advice), wise owls, companionable wolves (in contrast to the Red
Riding Hood wolf in the granny-suit, or Three-Little-Pigs wolves).
Later she interacts in much more depth, changing the rules for
herself as her succession of boyfriends progressively annoys her
brothers and alienates her parents — or whatever. She chooses
which parts of her culture she interacts with; going to work
or college, exploring sex or exploiting the possibility of sex as
power, and choosing her roles with respect to bosses, boyfriends,
mother. Her trajectory is complicit with that of her parents, her
friends, her work, as with all of us. Stewart and I like the model of
her intelligence interacting complicitly with the extelligence [16]
which surrounds her. This epigenetic, interactional-and-recursive

model for development of Maureen is acceptable to most people. To what extent is the model used to define Maureen for the rest of us who are not naïve DNA preformationists, asking if she has "the gene for" homosexuality or breast cancer? Maureen has many axes to develop around.

The model I offer is a bit complex for a novel, in which we usually see Maureen change along only one axis—she develops responsibility for her child, or she discovers her intellect, or she leaves the vapid but sexy Oscar for the deeply serious Arthur, or ... Only a few, rare medical experts see Maureen as such a complex structure—usually single-axis questions are asked by the tests, single-axis remedies proposed. A few endocrinologists and gynecologists do understand that there are many ways of being an effective endocrine woman, and don't simply look at her hormone levels and compare them with "the mean" [17], (compare, for example, [18]). More psychologists are prepared to see her as having her own solutions to, for example, her sexual problems or opportunities (although this may be changing as "genes for homosexuality" come to have currency in clinical psychology, and are being adopted into folk medical myths [19]).

But most papers in the field, many of the best books [1,15], most of the successful laboratories, promote and use genes-as-information and genes-mapping-to-single-characters metaphors. Those are the stories we find in the media: Maureen is her DNA-string made flesh. And science-fiction stories have run with this idea too, because it doesn't require much from the reader: "development" is not involved! Herbert's *Dune* series [20] is just such a model; the spice traders, the Bene Gesserit, all the characters are like chess pieces: they have their own moves, they do not develop. And the whole series only "develops" in the sense of a Greek tragedy or a chess game—not the kind of development a chicken or a Maureen exhibits, but only the development of a print from an exposed film. The picture is implicit in the system from the beginning. Like the pre-formed homunculus, it requires only to be expanded into space.

Nearly all embryos lose the game, of course, and that is another major problem with naive developmental metaphors [21]. Preformationist metaphors of development see reproduction as replication into the future. But it isn't like that in the real world: very nearly all sexually-produced creatures fail to breed, and nearly all fail to attain breeding age. Very nearly all spermatozoa fail to fertilize, nearly all fertilized eggs don't become breeders in their turn; this contrasts with Woody Allen sperms, Disney wild-life films which tell the romantic, fairy-tale version which does not require that viewers question the developmental process. On average in the real world, only two-ish offspring breed from each pair of parents, whether they lay 16 eggs in their lives like starlings or forty million eggs like cod. The Disney wild-life story has each little cuddly animal growing up; its model has every boot-black becoming a millionaire, like those 1920's films, and we can avoid thinking about all the ordinary folks — and millionaires, I guess — who became boot-blacks! Then we can enjoy the reproduction of organisms as a succession of preformations, with some accidents. All those baby starlings, dying young by a combination of suffocation and digestion as they're eaten, are unfortunate, not integral to our romantic view of Life in The Wild (which I'm sure we get, in the nursery, as part of our modern Western Make-a-Human-Being Kit with its roots in Rousseau). In contrast, those of us who have learned to see ecology, evolution and development as processes [21] find the real business of life more comprehensible, even if rather less comfortable.

Let's invent new metaphors, new stories in which the characters really do develop. Let us see Maureen as the complex she is, the creature she has helped make herself, not as a patient with BRCA1, or a gene for ... We are human beings. Let us discuss with her what she is, and what she wants to become.

References

[1] Dawkins, R. 1995 *River Out Of Eden*. New York: Basic Books.

[2] Nilsson, D. E. and Pelger, S. 1994 "A pessimistic estimate of the time required for an eye to evolve." *Proc. Roy. Soc. B.* 256 53–58.

[3] Gould, S. J. 1985 *The Flamingo's Smile*. New York: Norton.

[4] Townsend, S. 1994 *The Secret Diary of Adrian Mole*. London, Methuen.

[5] Milne, A. A. 1994 *Winnie the Pooh*. London: Methuen.

[6] Kuhn, T. S. 1973 *The Structure of Scientific Revolutions*. Chicago: International Encyclopedia of Unified Science.

[7] Eldredge, N. and Gould, S. J. 1972 "Punctuated equilibria: an alternative to phyletic gradualism," In *Models in Palaeobiology* (T. J. M. Schopf, ed.) San Francisco: Freeman, Cooper, 82–115.

[8] Cohen, J. 1996 "Who do we blame for what we are?" In *How Things Are: a science tool-kit for the mind*. Ed. J. Brockman and K. Matson. New York: William Morrow and Co, Inc.

[9] Cohen, J. and Stewart, I. 1991 "The information in your hand." *The Mathematical Intelligencer,* (13) 12–15.

[10] Cohen, J. and Rice, Sean H. 1996. "Where do biochemical pathways lead?" In *Integrative Approaches to Molecular Biology* (ed J. Collado-Vides, B. Magasanik and T. F. Smith). Cambridge, Mass: MIT Press. 239–51.

[11] Kauffman, S. A. 1993. *The Origins of Order.* New York: Oxford University Press.

[12] Waddington, C. H. 1962. *New Patterns in Genetics and Development.* New York: Columbia University Press.

[13] Goodwin, B. C. 1990. "Structuralism in biology." *Sci. Prog.* (74) 227–284.

[14] Cohen, J. and Stewart, I. 1994. *The Collapse of Chaos.* New York: Penguin, Viking.

[15] Dawkins, R. 1986 *The Blind Watchmaker.* Harlow, Essex: Longman

[16] Cohen, J. and Stewart, I. 1997 *Figments of Reality.* Cambridge: Cambridge University Press.

[17] Cohen, J. 1991. "Advances in Cell Biology." In *Scientific Foundations of Obstetrics and Gynecology* (eds. E. E. Philipp and M. E. Setchell) 4th Edn. London: Heinemann Medical Books. 1–18.

[18] O'Herlihy C. and Costigan D. C. 1994 "Female Puberty: normal and abnormal development." In *Understanding Common Disorders in Reproductive Endocrinology* (eds. M. M. Dooley and M. P. Brincat). Chichester: John Wiley and Sons Ltd. 23–33.

[19] Kaplan, L. J. and Tong R. 1994 *Controlling our Reproductive Destiny.* Cambridge, Mass: The MIT Press.

[20] Herbert, F. 1964 *Dune.* London: Gollancz Ltd.

[21] Cohen, J. 1996 "Reproductive fallacies." *Proc. Roy. Inst. G. B.* (67) 171–192.

6

ALGORITHMIC AND ASCETIC STORYTELLING: ALTERNATIVE APPROACHES TO IMAGINATION AND REALITY

Per-A. Johansson

> Reason uses imagination as a vestment outside and around it; if reason becomes too pleased with its dress, imagination, this imagination adheres to it like a skin; separation is effected only with great pain.
>
> *Hugo of St. Victor (d.1141)* [1]

> Rather than projecting the shadow of Tomorrow's unknowable realities, I propose to ask whether it is imaginable — I stress this crucial word — to exercise effective control over the future-shaping forces of Today. This rescues us from the impossible attempt to predict the shape of Tomorrow, and leaves us with the somewhat less futile effort of inquiring into the possibilities of changing or controlling the trends of the present. ...
> Can [the] intrusion of science and technology be bounded, confined to its needed applications, and kept from sucking the life out of our engagement with nature and with one another? I find that difficult to imagine.
>
> *Robert Heilbroner* (1995: 95,99)

... it is rather more difficult to recapture directness and simplicity than to advance in the direction of ever more sophistication and complexity. Any third-rate engineer or researcher can increase complexity; but it takes a certain flair of real insight to make things simple again.

E. F. Schumacher (1978: 150)

Preamble: Education and Imagination

When I first read Frank Herbert's *Dune* trilogy, some twenty years ago, I was particularly struck by his characterization of prophecy, and of the prophet's psychological quandary, spilling over into society at large. In one of the fictional quotes heading each chapter, some of the deeper issues involved are summarized like this:

Muad'Dib [the prophet featured in the book] gave us a particular kind of knowledge about prophetic insight and its influence upon events which are ... set to occur in a related system which the prophet reveals and interprets. ... such insight operates as a peculiar trap for the prophet himself. He can become the victim of what he knows ... The danger is that those who predict real events may overlook the polarizing effect brought about by overindulgence in their own truth. They tend to forget that nothing ... can exist without its opposite being present [2].

Education is founded on attempts at prediction. Our educational authorities try to predict what kinds of knowledge and what kinds of learning processes are likely to serve the future interests of students, and of the country at large. Increasingly, "serving our country" by means of education has come to mean "serving our country's economy," the meaning of "economy" being restricted to include profitable production and finance only, all other matters being made subservient to these. And since the increasingly global economies of all "developed" countries are squarely based on scientific and technological progress, science and technology, together with technical economics, naturally assume center stage in all educational efforts as well.

This education of ours, with its deep historical roots, and without which none of our technological and economic marvels could exist or continue to do so, is really all about one thing: it is *to prepare the imagination, by means of knowledge.* From a cultural point of view, the societal aim of education in science, technology and economics, is to establish *the conditions of intellectual believability and imaginability* that are conducive to the furthering of the imagined future position of ourselves and of our countries in the world economy. The measure of man these days is provided by interpreting the intricate results of the ever present, and typically modern, practice of double-entry bookkeeping. [3]

This actual state of affairs means that some things have become more believable and easier to imagine than other things, and it's all expressed in terms of knowledge. And it *is* knowledge. But it's not neutral; it's soaked in a certain vision of the world, whether this is acknowledged or not. This vision is expressed, if at all, in the form of stories. One such story about the past and about the future now appearing is "the algorithm story" I focus on below. Here's where Frank Herbert's warning about prophecy becomes relevant. The prophecy of the enthusiasts of our hi-tech civilization—the algorithm story, no less—amounts to, first, the symbiotic coevolution of humans with machines, and, second, the gradual replacement of humans by machines. Thus, it seems there's a real possibility of us becoming victims of what we know, by means of overindulging in our own truth. [4]

To my mind, the difficult and a little frightening thing to grasp about this, is that our technology, our science, and our economic system are both real and imagined *at the same time,* and more often than not it seems like imagination—embodied in education—comes first. Then, after having indulged in our creation, we'll have to live with the consequences. In order to bring home the depth of our present indulgence, I'd like to present a face to face confrontation of one current cosmogony (story of how the world came to be) with one *very* different kind of imagined and real world. This confrontation can be seen as a kind of serious mind-play; an exercise for the soul, the ancients would have called it. I think the contrast that will become apparent can tell us something important about ourselves or, more precisely, about what kind of intellectual and imaginative overindulgence that *we* are typically victims of.

Writing for the executives of the-world-to-be, James Martin

(1996: 6) summarizes the import of the information technology revolution for businesses everywhere like this:

> The entire nature of enterprises and employment is at the start of a massive historical transformation. The Internet enables chain reactions of technology feeding technology and people stimulating people worldwide. In aggregate it spells a momentous change that will affect the whole planet. The cybercorp [5] world is characterized by intense competition bypassing national frontiers and unions. It is spanned by dynamic computerized relationships among corporations using worldwide networks, with electronic reaction times, virtual operations, and massive automation ... In such an environment we all need to concentrate on how we delight our customers. How can we understand customer needs better? How do we eliminate defects? How do we cut costs? How do we improve quality? How do we make exciting products? How can we climb learning curves more rapidly even though complexity is higher?

The educational demands put by these developments have been well expressed, and put in their proper historical perspective, by James Bailey in his book *After Thought* (1996: 12):

> Even as tasks that used to belong to our minds are reassigned to electronic circuits and then evolve into completely new forms, we have the responsibility to be active partners in the new world that emerges. None of this is very far away in time. The individual building blocks of bit evolution are mostly in place. Computer scientists have been breeding and evolving programs successfully for over a decade. The techniques are increasingly well understood. Ultimately, it may take millions of computers evolving their programs based on trillions of pieces of information to be successful at large problems, but a million computers is no longer a large number. Nor in the age of satellites does a trillion describe a large number of pieces of information. Whether we prepare them for it or not, children born today will grow up to live in a world where computers outnumber them, where bit evolution plays a permanent role, and where [human] thought no longer holds the exclusive franchise.

Is this a world fit for human beings, one may wonder.

The Algorithm Story

The historical origins of the digital electronic computer are diverse, but intellectually one particularly seminal idea was Alan Turing's insight, that it's possible to construct an abstract automaton which acts as a *universal* computer [6] — "a general method in which all truths of the reason would be reduced to a kind of calculation" (Leibniz, *De Arte Combinatorica*, 1666). This rather seamless juxtaposition of the mathematician Turing of the 1930's with the philosopher von Leibniz of the 1660's hints at the deep historical sources of the wave of information technology now sweeping us all up on its crest, or dragging us along in its wake. With the computer, we've managed to represent the *logical processes* of our laboriously developed rational thinking in a general-purpose artifact, and we have quickly discovered that the potential of this intellectual prosthesis goes far beyond simple logic and arithmetic.

Already the pioneers of digital electronic computing (notably John von Neumann) were aware of the relevance of what came to be called "information theory" to the study of life itself. Von Neumann's ideas about self-reproducing automata (posthumously published in 1966) were seminally confirmed, it seems, by the discovery of the molecular structure of DNA, and of the genetic code, by means of which the genes "inscribed" in DNA (which are inherited) are "transcribed" into proteins (which build up and maintain the new body). This chemical process, "the *stuff* of life," could meaningfully be expressed and studied in the language of information theory. With such intellectual building-blocks in place, it was inevitable that speculatively inclined scientists would begin to envisage the origin and history of the living world in terms of an appropriately fashioned algorithmic theory. Fortunately, one was available already. It just had to be reformulated in the new idiom — or so the proponents of the new view proclaim, not without a catching enthusiasm.

To make a quite complex and ambiguous story very short indeed, it turned out — with full force only since the 1980's — that the Darwinian theory of evolution by means of natural selection could be conveniently rephrased as being a matter of algorithmic *processes*, i.e., of the physical, four-dimensional time-space execution of formal recipes for "sorting, winnowing and building things" (Dennett 1996: 52). These kinds of evolutionary algorithms do not necessarily satisfy the usual strictness of mathematical criteria (provable, guaranteed to terminate) and, because of this, the new field of evolutionary computation is quite messy, compared to the stark and clear light cast by classical mathematical science. Still, the possibility of using computers for running (simulating) evolutionary algorithms — largely, but not exclusively, culled from Darwinian theory — has thrown an altogether different light on an old conundrum confronted by evolutionists:

> Most of the controversies about Darwinism ... boil down to disagreements about just how powerful certain postulated evolutionary processes are — could they actually do all this or all that in the time available? These are typically investigations into what an evolutionary algorithm *might* produce, or *could* produce, or is *likely* to produce, and only indirectly into what such an algorithm would *inevitably* produce (Dennett 1996: 57).

I think it's clear that the patterns discerned by theoretical ecologists in real ecologies in nature *look* uncannily like the artificial ecologies "grown" (like Thomas Ray's *Tierra*) inside computers, using the same "methods" nature is supposed to do. So, these evolutionary algorithms *really can* produce (in computers, that is) a continually evolving, emergent complexity. This complexity certainly resembles "life," thus giving the algorithmic approach to evolution a further, very powerful, rhetorical boost — given the ubiquity and familiarity of computers nowadays.

It's probably a safe bet that much of biology, economics and also sociology (cf. Epstein and Axtell 1996) will be replete with these new "intermaths" (Bailey 1996) [7] in the future. In this way we and our world can be envisaged and studied, more rationally than ever it seems, as having arisen and evolved quite automatically and unconsciously, at all levels. [8] Even human co-operation and the "social contract" can well be studied by means of computer-simulated abstract games, evolutionarily interpreted (Axelrod 1990, Skyrms 1996). And no one seems to find this very strange.

All of this means that simultaneously with the growth of an industrial *economy* based on information technology (Castells 1996), the evolution of *life* has turned out to be a kind of informational engineering process too, at least in the minds of what I will call the "algorithmic storytellers." The story they tell heralds, in a sense, the ascendancy of thought — not necessarily human thought — over matter, since anything capable of being described (or evolved) recipe fashion by means of some programming language can actually be implemented on computers. It seems thoroughly appropriate, then, that the new cosmogony now in the making tells us that the world itself, ourselves included, *is* such an implementation. This promising infant worldview is "the algorithm story."

A characteristically eloquent quotation will give us the flavor of it — Darwinian biologist Richard Dawkins (1988: 111):

> It is raining DNA outside. On the bank of the Oxford canal at the bottom of my garden is a large willow tree, and it is pumping downy seeds into the air. There is no consistent air movement, and the seeds are drifting outwards in all directions from the tree. Up and down the canal, as far as my binoculars can reach, the water is white with floating cottony flecks, and we can be sure that they have carpeted the ground to much the same radius in other directions too. The cotton wool is mostly made of cellulose, and it dwarfs the tiny capsule that contains the DNA, the genetic information. The DNA content must be a small proportion of the total, so why did I say that it was raining DNA rather than cellulose? The answer is that it is the DNA that matters. The cellulose fluff, although more bulky, is just a parachute, to be discarded. The whole performance, cottonwool, catkins, tree and all, is an aid of one thing and one thing only, the spreading of DNA around the countryside. Not just any DNA, but DNA whose coded characters spell out specific instructions for building willow trees that will shed a new generation of downy seeds. Those fluffy specks are, literally, spreading instructions for making themselves. They are there because their ancestors succeeded in doing the same. It is raining instructions out there; it's raining programs; it's raining tree-growing, fluff-spreading, algorithms. That is not a metaphor, it is the plain truth. It couldn't be any plainer if it were raining floppy discs.

This is the story, *the plain truth* of the information society: "what we [all living creatures] actually do is exactly what a [self-duplicating von Neumann machine] [10] is defined as doing. We roam the world looking for the raw materials needed to assemble the parts needed to maintain ourselves and eventually assemble another robot capable of the same feats. Those raw materials are molecules which we mine from the rich seam of food" (Dawkins 1997: 258). As has been made clear by Daniel Dennett and others, the evolution and functioning of the living world, and the evolution and functioning of the human brain-mind (including its artifactual products) can profitably be seen as two aspects of the same general history. In theoretical terms: *"there is only one Design Space, and everything actual in it is united with everything else"* (Dennett 1996: 135; emphasis in original). When it comes to the mind, this position has been succinctly expressed by Eric Dietrich (1994: 14):

> All of our cognitive processes, from seeing, hearing, walk-ing, and tasting to reading, writing, learning, reasoning, etc., are the computation of some function. And, going the other way, anything that computes these functions will think (see, hear, walk, taste, learn, write, reason, etc.). This view is called *computationalism*. Specifically, we all believe that explaining human (and any other animal's) cognitive capacities requires discovering which functions their brains compute and how they compute them (i.e., both which algorithms they use and how the algorithms are implemented in the neural hardware). And to build an intelligent, thinking computer, we need only program it to compute these (or relevantly similar) functions. However, we don't necessarily have to implement on the ma-chine the very same algorithms humans use (remember, there are many algorithms for a given function).

The Achilles Heel

This emerging algorithmic view is strongly coherent, in ratio-nal terms. To my mind, though, the crux of the matter is that a *really* coherent worldview must also include a spiritual and moral teaching and practice, as well as a scientific description and explanation of the physical world. This is not yet the case in relation to our dependence on the computer. This "root arti-fact" of the view discussed here, which is also its root metaphor, seems to embody infinite possibilities, and this is the basis for the fascination of the story of the world told by means of it. Its enthusiastic proponents are characteristically defiant in the face of all opposition, real or imagined. Daniel Dennett (1996: 451) expresses the pose in a nutshell:

> It is time to turn the burden of proof around .... Those who think the human mind is nonalgorithmic should consider the hubris presupposed by that conviction. If Darwin's dangerous idea is right, an algorithmic process is powerful enough to design a nightingale and a tree. Should it be that much harder for an algorithmic process to write an ode to a nightingale or a poem as lovely as a tree? Surely ... [e]volution is cleverer than you are.

There's no possibility of misinterpreting this: if you think you're *not* an algorithmic process, you're dead wrong. Well, I'm certain I'm not, or, more precisely, I'm certain that I'm *not only* an algorithmic process. This conviction is based on an intrinsic aspect of *human* life that is usually ignored by this brand of thinkers: the proof of a pudding is in the eating. To this hoary wisdom I would like to append a footnote: "eating" does not equal "technological implementation." (By "pudding" I mean our life together, as construed by our telling each other stories of who we are.) One algorithmic story-teller, however, that seems to realize this, when he thinks about the matter of *doing good and not evil*, is Dennett himself. This provides us with an opening that makes possible a more than superficial confrontation of today's really hot science story with a more ancient wisdom, one consciously concerned with a change of heart rather than with knowing everything. I'll come to the latter, ancient, tradition in a moment.

The strands of the algorithmic narrative intrude into our most cherished inherited notions, like that of "life" itself, not to speak of the meaning of "human." We may note especially the metaphorically illuminating (but also, I think, distorting) use of engineering concepts like "design," "robot," "implementation" etc. It's important to understand that this, in the current climate of opinion, does *not* imply a mechanistic worldview in the 18th century sense. The old distinction between "mechanistic" and "organic" ontologies (cf. Merchant 1983) has recently broken down almost completely (see, e.g., Channell 1991, Emmeche 1994, Keller 1995, Kelly 1995). Even so, it's noteworthy that what used to be the mysteries of life can now be approached as a series of (software) engineering problems or, more precisely, since it's all about evolution, reverse engineering problems. Once a successful reverse engineering analysis has been accomplished, it also becomes possible — thanks to the open-endedness of computer programming — to *forward*-engineer products embodying the discovered principles (if not the actual processes). This aspect of the story, I think, is particularly important to highlight, since it means that it's not just a question of discovering the secrets of creation, but also of literally creating our own future.

The Ascetic Story

Now, for purposes of obtaining an alternative perspective on these intellectual and imaginative developments, I would like to introduce the reader to a *quite* different narrative of man and his brain-mind. The two stories are equals, culturally speaking, but very much unidentical, substantially. I've chosen to treat them as comparable, however, and the leading question in all that follows is: If the ascetic story outlined below is taken seriously, what would its verdict on the algorithmic story be? I think it's very illuminating to put this question as forcefully as possible.

In another era, in another society (albeit surviving today as a relict), *the plain truth* was different, and the story too. To get the flavor of it, we may quote St. Neilos the Ascetic (*d.* ca 430). Please read it very carefully a couple of times, with the footnotes. The conceptual meanings, although strange to us, are very exact.

> The story of Ish-bosheth [cf. 2 *Samuel* 4:5-8] teaches us not to be over-anxious about bodily things, and not to rely on the senses to protect us. He was a king who went to rest in his chamber, leaving a woman as door-keeper. When the men of Rechab came, they found her dozing off as she was winnowing wheat; so, escaping her notice, they slipped in and slew Ish-bosheth while he was asleep .... Now when bodily concerns predominate, everything in man is asleep: the intellect, the soul and the senses. For the woman at the door winnowing wheat indicates the state of one whose reason is closely absorbed in physical things and trying with persistent efforts to purify them. It is clear that this story in Scripture should not be taken literally. For how could a king have a woman as door-keeper, when he ought properly to be guarded by a troop of soldiers, and to have round him a large body of attendants? ... But

improbable details are often included in a story because of the deeper truth they signify. Thus the intellect [11] in each of us resides within like a king, while the reason [12] acts as door-keeper of the senses. When the reason occupies itself with bodily things—and to winnow wheat is something bodily—the enemy without difficulty slips past unnoticed and slays the intellect. ... What advantage do we gain in life from all our useless toil over worldly things? ... Through our anxiety about worldly things we hinder the soul from enjoying divine blessings and we bestow on the flesh greater care and comfort than are good for it. We nourish it with what is harmful and thus make it an adversary, so that it not only wavers in battle but, because of over-indulgence, it fights vigorously against the soul, seeking honours and rewards.

If this quotation (from *Philokalia,* [15] I: 210-211) seems very different from, and somehow at odds with, the one about a "rain of instructions" in Oxford, I agree. The contrast is deliberate. It illustrates the mental chasm separating the best minds of our culture from those of another. What should make one stop and think, is that both types of story-telling, in their origins, are deeply human (although the human potentialities expressed are widely dissimilar), and that both are equally metaphorical—Richard Dawkins' disclaimer notwithstanding. Interestingly, St. Neilos seems to be much more aware of this than Mr Dawkins. The latter's "improbable detail" (the rain of instructions) is regarded as literally true, while the former's improbability (a lone woman acting as door-keeper to a king) is seen as a metaphor superficially hiding a deeper insight. St. Neilos' culture is fully aware of the dark depths of the human heart, and is out to cure it; Mr Dawkins' culture is busy unravelling the secrets of creation, and is out to implement its knowledge in artificial products and even creatures.

The societal context of the algorithm story is clear enough, since we're living right in the middle of it. But what's the broader context of the ascetic story? The Eastern Orthodox Church [16], of which St. Neilos is a representative, is alive and well at this moment, too, but its storytelling stands in a very different relationship to society and the economy at large, now, than it did in former days. This is evident if we turn our minds to Byzantium, where Orthodox culture had its first, longest and most luxuriant flowering. When hearing talk of monks and nuns today, most readers will probably think of men and women celibates living secluded from the rest of society in monasteries and convents. However, this train of associations is profoundly misleading in the context of this essay and is best forgotten. The whole point of the confrontation I will set up below is lost if that chain of thought is persevered in. Hopefully this quote from historian Rosemary Morris (1995: 1) will set the reader's mind off in a somewhat different direction:

> [M]onks did not constitute a separate caste within Byzantine society. They might follow different ways of life, or adhere to different spiritual priorities, but monks had all once been laymen and many laymen, after long years in the secular world, became monks. "Abandoning the world" thus often meant not the abandonment of human relationships such as family feeling or friendship, or the discarding of claims to leadership in society, but the recasting of them in a different, spiritually orientated context.

The important thing to grasp is that monks (and nuns) were not a breed apart, and didn't regard themselves as such, at least not habitually and commonly. Spiritually motivated ascesis was an ideal familiar to anyone (and still is, among the Orthodox) [17]. The main differences between monastics and laymen were (are), first, an "inverted" sense of priorities (the monastics being firmly *committed* to spiritual and moral repentance and devotion to God), and, second, that the monastics practiced an age-old *discipline* of prayer and spiritual combat. But the values espoused and the virtues practiced by monks and nuns *permeated the rest of society* as well, just like computers and their antics permeate our society, together with an experience of rampant societal change — evolution! Consequently, the societal value and "position" of ascetic storytelling in Byzantine society (if we stick to that as prototypical), is very closely analogous to the value and position of algorithmic storytelling in ours. We'll come back to these contextual matters later on. Now we'll return to sampling the ascetic story itself.

The story of the world as told by the Orthodox ascetics, is reminiscent in significant ways of the modern evolution story — whether algorithmic or not: There's a "beginning;" there's a "force" (the *Logos,* natural selection) guiding the appearance of everything, and in both cases man, in some unfathomable fashion, seems able to "sum it all up" in his very being. The difference is one of emphasis, but also of the meaning of "beginning" or "origin." Kallistos Ware, an Orthodox bishop and monk living in Britain, writes:

> In saying that God [18] is Creator of the world, we do not mean merely that he set things in motion by an initial act "at the beginning," after which they go on functioning by themselves. God is not just a cosmic clockmaker, who winds up the machinery and then leaves it to keep ticking on its own. On the contrary, creation is *continual.* If we are to be accurate when speaking of creation, we should use not the past tense but the continuous present. We should say, not "God made the world, and me in it," but "God is *making* the world, and me in it, here and now, at this moment and always." Creation is not an event in the past, but a relationship in the present. If God did not continue to exert his creative will at every moment, the universe would immediately lapse into non-being; nothing could exist for a single second if God did not will it to be (Ware 1979: 57).

What is the connection between this view of *creation* with the view of *human psychology* hinted at by St. Neilos above? The key is the Orthodox doctrine of man as created "in the image" of God. "Man's unique position in the cosmos is indicated above all by the fact that he is made "in the image and likeness" of God (*Genesis* 1:26). Man is a finite expression of God's infinite self-expression" (Ware 1979: 64–65). Thus, whatever man conceives of and does, is possible only because of his God-given participation in the ongoing creation, one might say:

> Fundamentally, the image of God in man denotes everything that distinguishes man from the animals, that makes him in the full and true sense a person — a moral agent capable of right and wrong, a spiritual subject endowed with inward freedom (ibid., 65). The divine is the determining element in our humanity; losing our sense of the divine, we also lose our sense of the human (ibid., 67).

The following passage from St. Maximos the Confessor (*d.* 662) ties these two strands — creation and psychology — still more firmly together:

> God, in whose essence created beings do not participate, but who wills that those capable of so doing shall participate in Him according to some other mode, never issues from the hiddenness of His essence; for even that mode according to which he wills to be participated in remains perpetually concealed from all men. Thus, just as God of his own will is participated in — the manner of His being known to Him alone — in the surpassing power of His goodness, he freely brings into existence participating beings, according to the principle which He alone understands. Therefore what has come into being by the will of Him who made it can never be coeternal with Him who willed it to exist. The divine Logos [see note 11], who once for all was born in the flesh, always in His compassion desires to be born in spirit in those who desire Him. He becomes an infant and moulds Himself in them through the virtues. He reveals as much of Himself as He knows the recipient can accept. In this way the divine Logos is eternally made manifest in different modes of participation, and yet remains eternally invisible to all in virtue of the surpassing nature of His hidden activity (*Philokalia*, II: 165-66).

According to this ancient doctrine, then, the important thing to emphasize is *that* man/the world depend on God. Neither the world nor humankind can exist on their own account. The question of *how* the world, and man as a participating agent in it, come to be as they are, physically speaking, is simply irrelevant to the concerns of the tellers of this story. What they *do* address, which the evolution story doesn't, is the question: What do we live *for?* And the Orthodox answer is: to help redeem the creation, first of all by spiritualizing our bodies by offering them to God (Ware 1979: 64). Thus man's "function" in the divine scheme is to embody the Spirit of God (technically, this is known as *theosis*), even though the kingdom of God is not of this world (1 *Corinthians* 15:50) and never will be (*Revelation* 20–21). This, of course, is completely beyond the bounds of algorithmic storytelling, but it's important to realize, also, that neither does it contradict it; the evolutionary story is subsumed, rather, and relativized where the total human reality is concerned.

The real confrontation between these views, then, doesn't take place in the realm of science, but on the stage of human social life. But, and this is a complication, the full force of the ascetic story can only be felt and responded to *if it's recognized to be true,* [19] i.e., if it's experienced as corresponding to a greater reality, indicated by the story itself. This circumstance will lead us, a little bit further on, to discuss the societal conditions of believability, i.e., the actual ambience in which decisions of this nature are really made. Even if the algorithmic and the ascetic stories don't address the same issues, there's a connection — namely us. In order to see the point of confronting them, one should avoid thinking in terms of "describing the world." Rather one should think of the nature of the different personal and societal *impacts* of the stories. It may be that they address different parts or abilities of the human body-mind, [20] relating to different (proto)types of human conscious awareness.

Confrontation, I

If we begin now to confront the two stories in earnest, we can say that the fact that Charles Darwin seems to have found a better technical explanation for the origin of the *physical design* of living things, than the Archdeacon William Paley (a respected target of algorithmic storyteller Richard Dawkins), *doesn't* mean that God isn't creating the world right now. If so, the assumption that "physical design," as such, equals "all there is and can be" is a matter of faith, not Darwinism. And in matters of faith the criteria for sound judgment are very different from what they are when competing scientific hypotheses are at stake. Richard Dawkins argues incisively against what he calls "the Argument from Personal Incredulity" (Dawkins 1988: 38f). What he doesn't mention (does he even notice it?) is that his own argument for universal Darwinism can as well be labeled an "Argument from Rational Credulity," i.e., a credulity based on reason and *nothing but* reason. It is in this connection, I think, that the criteria for comprehensive intellectual judgment (including the question of what to put one's faith in) must be discussed critically. It is one thing to argue, on scientific grounds, that the existence of eyes (traditionally a tough nut to crack for Darwinists) *can* plausibly be explained in Darwinian terms (Dawkins 1997: 126-179). It is quite another to state confidently that we no longer have to resort to superstition when faced with the deep problems: Is there a meaning to life? What are we for? What is man? After posing the last of these questions, the eminent zoologist G.G. Simpson put it thus: "The point I want to make now is that all attempts to answer that question before 1859 are worthless and that we will be better off if we ignore them completely" (Dawkins 1976: 1).

What I'm saying, what this whole essay in a sense is about, is that Dawkins *et alii* [22] argue against a *misconstrued* Christian doctrine, something they generally do very well. Their line of attack is understandable, given the fact that the misconstruction, in important respects, is not theirs, but rather that of their contemporary Christian opponents—largely made up of various Protestants, more or less literalist in outlook. This means that something historically and intellectually important is habitually left out of the discussion, making it more superficial than it could—or should—be. The issue is whether the same basic criteria really are applicable to questions that may concern very different modes of reality or, perhaps better, modes of humanly participating in reality.

It's too seldom noted (as Bishop Ware does in the quotation above), that the fathers of the Christian doctrine of creation don't answer the same set of questions as Darwinism does. In particular, the Christian doctrine *isn't concerned with origins in a historical sense.* The Orthodox fathers' sense of "history" is completely different from Darwin's, and the two senses cannot be compared directly. This means that on the rational level the problem is definitely *not* a question of "choosing" between clearcut alternatives, but rather one of trying to establish where they actually meet [23], and where they truly differ.

Many critics claim that the algorithmic story (including Darwinism in its current sense) is inhuman and dehumanizing in its implications, while its advocates see it as a long-awaited extension and magnification of our truly human potential of understanding reality, based on our knowledge of the physical world. The classical Christian stance doesn't entail choosing sides in this manner. Its position is that there are in us two wills (one worldly, one divine), the first of which should be subjected to the latter—as frequently emphasized by the revered apostle to the gentiles, St. Paul. He also observed in this connection that "all things are lawful, but not all things are helpful. All things are lawful, but not all things build up. Let no one seek his own good, but the good of his neighbor" (1 *Corinthians* 10: 23-24). I think that these statements, *and* their theocentric context, are far from irrelevant to our present- day concerns. Technological innovation and scientific knowledge are not condemned, on this view, but they're dethroned and put in the proper place in relation to the real business of human life, which isn't business, nor unbridled curiosity for its own sake and regardless of consequences.

Imagination

What's at issue, as I see it, is not only the way we "narrate the world" to each other—in order to situate ourselves, our activities, and our dreams in an imaginatively coherent context—but also the more or less *implicit* values incorporated into any story, whether scientific or not. I think it's extremely important, a matter of life and death even, to begin to understand explicitly the nature and import of our capability to actively *imagine a reality* and then, step by step, to *realize it physically*. The reason this is serious business, of course, is that the products of our imagination are not all life-enhancing in some way; many are deadly (if not for the body, then perhaps for the soul), and it can be very difficult to decide in advance which is which. Is such a judgment even possible? This is a question I think it would be very good if we could become able to answer in some fashion. By "we" here I mean especially scientists, technologists, businessmen, economists, and politicians—all those directly and actively involved in imagining and implementing *what will become* everyone's future, in one way or another. Strictly speaking, it's a matter of personal responsiblity, no less.

Recently, The MIT Press issued a fascinating volume called *HAL's Legacy: 2001's Computer as Dream and Reality,* edited by David G. Stork of the Ricoh California Research Center. What this book makes clear, sometimes explicitly, sometimes between the lines, is that the sharing of a certain vision of the future may result in very definite artistic, scientific and technological achievements, all related to a common imaginative core and, furthermore, that the artistic articulation of the vision predates its actual implementation. No one has yet been able to build HAL in reality, to be sure, but the dream made explicit in Arthur C.

Clarke's and Stanley Kubrick's movie is very much alive, and it motivates scores of competent workers. And results are certainly forthcoming; whether these are really "HALish" or not is to some extent beside the point. The important question is: What happens when *this kind of dreaming* is married to the proven efficacy of modern science and technology? Before long, much of it won't be just a matter of cinematic fantasy any longer. Where will we, the dreamers, be then? Do we *have* to use our creative imagination in this way? Should we perhaps feel responsible, in some way, not only for what we actually do, but also for what we collectively imagine? Can there be such a thing as *"pre*sponsibility"?

What distinguishes the fertility of current *scientific* imagination from the ordinary garden variety, is that it's constrained within a time-honored tradition incorporating highly disciplined modes of reasoning, experimentation, observation, mutual criticism and support, and, of course, the established theoretical and practical results of previous research. To some, this connection between constraints and creative fertility may seem to be a contradiction in terms, but this is in appearance only. The facts of the matter can be metaphorically illustrated (or should I say simulated?) by looking at the possibilities inherent in the use of formal programming languages. These languages are syntactically heavily constrained, to say the least, and yet they can be made to evoke (I know no better word) almost anything. According to some, they may even be used to create life — quite literally (Langton 1996).

There's also, in fact, a very close connection between science and computation, as evidenced by this definition of what a *scientific theory* is:

> [A] scientific answer to a question takes the form of a set of rules or a program. We simply feed the question into the rules, turn the crank of logical deduction and wait for the answer to appear as the output of program. Of course, not just any set of rules will do. In order for the rules to be scientific, they must pass certain tests of reliability, objectivity, explicitness, public availability, and so forth ... , but once such filters have been applied, what remains is pretty much an algorithmic notion (Casti 1997: 197).

What are we to make of this in relation to our imaginative capabilities? To make a long story short, I think it means that we (i.e., "modern man") have managed to devise *an extremely powerful way of realizing the fruits of our imagination.* This makes it necessary to discuss it from a spiritual and moral perspective as well, focusing particularly on the *aims and ambitions* occupying the imagination of the practitioners of science and technology, and also that of their receivers and consumers.

As I've noted several times already, the ultimate goal of the algorithmic imagination now rapidly gaining ground, is not only to describe and explain evolution, but also—or even primarily—to implement its findings in artifacts, the more autonomous the better:

> I wish to build completely autonomous mobile agents that co-exist in the world with humans, and are seen by those humans as intelligent beings in their own right. I will call such agents Creatures. This is my intellectual motivation. I have no particular interest in demonstrating how human beings work, although humans, like other animals, are interesting objects of study in this endeavor as they are successful autonomous agents. I have no particular interest in applications; it seems clear to me that if my goals can be met then the range of applications for such Creatures will be limited only by our (or their) imaginations. I have no particular interest in the philosophical implications of Creatures, although clearly there will be significant implications.

These symptomatic statements were made by Rodney Brooks [24], a well-known and influential roboticist at MIT, originator of the so-called "subsumption architecture," which, to put it simply, entails that the robot, in a way, "learns" how to conduct itself in the world, rather than having to be programmed by a human programmer. What amounts to the same principles, intellectually, characterize what James Bailey calls the new "intermaths"—self-evolving and -learning software agents, processed in parallel (virtually or actually) rather than sequentially. It's these scientific developments, closely allied with the Darwinian evolution story [25], that form the wherewithal of many computer-based imaginings today.

What was (or, rather, is) the goal of the Orthodox ascetic imagination? In a sense its goal too is "implementation," but not in the form of autonomous artifacts, but in a change of heart — technically known as *metanoia*. This term is traditionally translated into English as "repentance." That it has a moral connotation is clear enough; what's generally not clear to the modern reader, is that this moral connotation substantially presupposes a change in man's *reason,* or, more precisely, in the ontological basis of his reasoning. Thus it isn't a matter of morals in the usual sense of the word, but has more to do with realizing one's proper intellectual and spiritual direction. This is plain in the original Greek [26]. It's not for nothing that the practitioners of the so-called hesychast [27] tradition in the Orthodox Church calls their foremost discipline or method noetic prayer — something closely akin to (but not identical with) what Buddhists mean by "meditation" (cf. Ware 1989). Prayer, as a concentration and opening of the mind, can be conceived of as directed *either* at God, *or* at what effectively functions as one's god. *Metanoia* results, among other things, in a different conscious relationship to the imaginative faculty of the mind: "Beware of the imagination *(phantasia)* in prayer — otherwise you may find that you have become a *phantastes* instead of a *hesychastes!*" (Ware 1989: 25, referring to St. Gregory of Sinai). Many creators and users of algorithmic stories and devices would do well to heed this warning — if only they were aware of its import. [28]

Confrontation, II

In order to call a somewhat different tune, I will now present a tabular comparison of some key items of these two very different human worlds (the computer-mediated current one and the *Logos*-mediated Byzantine one). The comparison is restricted to the intellectual elites of the respective cultures (even though the meaning of "intellectual" itself is fundamentally different in the two contexts). "Monastic culture" should be interpreted as that body of practice, thought and artifacts lived and created by the very best minds of Orthodox Christian culture; in this connection the reader is reminded that the monastic ethos is (or rather was) *intrinsic to all of society* in traditionally Orthodox lands. "Computer culture" should be interpreted in a corresponding fashion; that its products, including the stories, are ubiquitous and essential to our societies (or at least economies) need not be emphasized further. The following table is largely a pedagogical exercise, and all its implications and ramifications cannot be pursued here [29]. It's best approached as a rhetorical imaginative tool, making a point that is to some extent beyond words.

Monastic Culture	Computer Culture
theocentric; quiet personal conformation to the Creator resulting in contemplative love (theoria), with divinization (theosis) as goal; not subject to historical change	**technocentric**; active, externally creative realization (implementation) of possibilities resulting in rationally constructed algorithms, with complex artifacts as goal; subject to accelerating historical change
basic reality: God, *Logos* (Christ); man as the image of God; loving wisdom	**basic reality:** quantified material units; binary logic; rational operations
central artifact: the Bible; the Liturgy	**central artifact:** the digital computer; software
method: noetic prayer; *lectio divina;* virtue (moral discipline, imitation of Christ); meditating on biblical themes and symbolic images (icons)	**method:** algorithmization; programming; rational (logical) discipline; technical accuracy; meditating on simulations
epistemological bias: understanding = introspective love (Christian *gnosis*); implicit personal experience and understanding	**epistemological bias:** understanding = explicit abstract and concrete (verbal and technological) constructions

Monastic Culture	Computer Culture
concept of truth: "truth" equals the true life, i.e., a life that is true to its divine origin and purpose	**concept of truth:** "truth" equals true statements, i.e., statements that can be shown to correspond to empirical findings and/or that can be implemented in functioning devices
communicative activities: personal relations within the confines of a community; inspiring sermons; edifying counseling; public readings; letters; manuscripts; walking, riding, boat travel	**communicative activities:** man-machine interaction; lectures; abstract reasoning and exchange of ideas in relation to machines (real or virtual); e-mail; printed matter; telephone; car driving, train travel, air travel
results: contemplation; the human embodiment of divine love; *locus:* one's own "heart" and, consequently, the community	**results:** explicit theoretical models; implementations; *locus:* the material, artifactual and social world
ambience: monasteries; Churches; divine services; relatively small manuscript libraries; farming and other manual crafts	**ambience:** computer laboratories; "cyberspace"; high-technology devices and appliances of all kinds; very large libraries of printed matter

Monastic Culture	Computer Culture
eschatology: participation in eternal life here and now; resurrection of the body in a new creation at the end of time	**eschatology:** limitless implementation of algorithms; autonomous thinking and living computers; infinitely extended cyborg existence

The thinking behind the tabular comparison goes as follows. Efficacious storytelling requires a responsive audience and an appropriate ambience. Neither algorithmic nor ascetic storytelling take place in a cultural vacuum. Many things must be unthinkingly presupposed in order for the stories to have an impact, or even to seem plausible at all. These preconditions have less to do with any abstract principles pertaining to argumentation and storyline, than with the actual cultural ambience in which the narration takes place. This should be understood very concretely. That's why it requires a real effort of our active imagination to grasp the import and meaning of the ascetic story, in contrast to the algorithmic, which is very much in tune with the times — or soon will be, since it's helping create the times we'll actually live in shortly.

The key to a meaningful philosophical and existential confrontation between science and technology, on the one hand, and Orthodox ascetic practice, on the other, is to look at them in terms of their respective relationships to *time*. Technological and scientific history is evolutionary in a Darwinian sense, i.e., current features and developments depend upon and/or outcompete older or alternative products or hypotheses [30]. This fact also conditions the vision of the future embraced by practitioners; the only thing certain is that the future will be as different from the present as the present is from the past. Evolution won't stop; if it"ll turn out any *better* is a different question altogether. Thus the "existential ambience" of modern science and technology is one of *incessant change,* essentially "out of control" (Kelly 1995). It's both in and of this evolving world, "and the world passeth away, and the lust thereof" (1 *John* 2:17a).

Ascetic spiritual practice has a very different conscious relationship to the world of change. It aims at continually, and quite bodily, transforming the relationship to it into one of contemplative perfection *(theoria)*. This process is the *theosis* mentioned before, which proceeds by means of personal realization; it does not issue from propositional decrees of any kind [31]. The past, the present, and the future — all human time — is thus viewed in terms of its relationship to the eternal Creator of all. This attitude to the world as such is summed up in the words of *Ecclesiastes* 1:9: "What has been is what will be, and what has been done is what will be done; and there is nothing new under the sun." Essentially, everything in the world of change is always the same, even though it is always different. This characteristic paradox is resolved if one realizes that whatever physical novelties evolution throws up, they are of *the same ontological kind*, metaphysically speaking, as everything that went before. (This, as I see it, is the true meaning of the universal Darwinism expounded by Richard Dawkins and Daniel Dennett.) Because of its attitude of non-involvement, in the sense of not being *passionately* involved in the world of change, the "existential ambience" of ancient ascetic theology is one of *calm and quiet*. It's in but *not of* this world, the repenting Christian having put himself under the guidance of an unchanging unifying principle; "he that doeth the will of God abideth for ever" (1 *John* 2:17b).

In these terms, the different biases of technology and of science, on the one hand, and of ascetic theology, on the other, can be seen to be completely logical in their consequences. The bias of the former implies the "necessity" of the implementation of results, in an atmosphere of creative competition, without anyone giving much thought to where it all might lead. The bias of the latter implies the necessity of bringing men and women into permanent contact with their deepest well-springs, to make the love of their eternal Creator shine forth in time. It's consciously disciplined in quite the opposite direction from the demands of our hi-tech economy. Ascetic practice and theology, then, can be seen as reflecting *another (proto)type of mind and consciousness* than man's scientific curiosity and technological creative urge. The real question is: what (proto)type shall one put one's faith in?

At one time, even in Western Europe, care for the soul was an intrinsic part of higher education, and knowledge of the world was not separated in thought from the wisdom of God (accessible to the human heart through prayer; and indirectly in other ways). This changed, however, due to an unprecedented series of contingent historical developments too varied and intricate to go into here. Whatever the events and causes, the severing of knowledge from loving wisdom — in theology, philosophy and science alike — has now spread throughout the world by force of circumstance. The crucial point of this essay, somewhat obliquely indicated, is that this state of affairs has less to do with purely intellectual difficulties and seeming inconsistencies, than with the actual and concrete circumstances of life in "developed" countries. The demands are so pressing, the time always so short, that only a few are even able to find the serenity necessary for focusing mentally in the proper direction. No wonder we're drowning in substitutes.

Reality

> Enlighten the eyes of our understanding and raise our mind
> from the heavy sleep of slothfulness.
>
> *St. Basil the Great (d.* 379) [32]

We can now reconnect with the wisdom of St. Paul, already quoted above; "All things are lawful, but not all things build up," he wrote. A little further on in the same letter he urges his readers to "be imitators of me, as I am of Christ" (1 *Corinthians* 11:1; cf note 11 on *Logos!*) — that is, in order for us really to be able to build up: "According to the grace of God given to me, like a skilled master builder I laid a foundation, and another man is building upon it. For no other foundation can any one lay than that which is laid, which is Jesus Christ" (1 *Corinthians* 3:10-11). This captures the inner truth of the entire Christian ascetic and intellectual tradition, but, when push comes to shove, is this still relevant to us post-Christian and generally irreligious moderns? I think it is, provided it is realized *in the face of* sophisticated contemporary scientific thought and technological innovation. (This realization, however, is primarily a matter of awareness, and not of any specific belief content; cf. note [33].)

Leaving all exegetical niceties aside, I would like to relate St. Paul's exhortation to the tabular comparison I've made between the world of our algorithmicists (and, increasingly, of us all) and the world of the ascetic theologians of the Eastern Church. The world *we will live in* is created right now, and I think that one key to an appropriate response to current trends lies in realizing the possibility of disciplining our imagination in the direction indicated by the Orthodox ascetics so long ago. If all we can imagine is more of the same ontological kind (even more change, even more things to sell), then what's the point? On the other hand, if we could but *learn* (or relearn) to imagine something completely different, who knows what would happen? [33]

Robert Heilbroner (quoted at the head of this essay) finds it *hard to imagine* us putting a boundary on the intrusion of technology and science into the innermost recesses of our personal lives. I would urge the reader to seriously consider the possibility, that the actual ambience of our lives in the modern world almost fatally *distorts our judgment* on these crucial issues. Our daily contributions to the smooth but hazardous functioning of the global economic artifice provokes and enthuses us into so much time-consuming effort, that we easily lose our awareness of any vestige of spiritual discrimination that we may still harbor, deep down. *Can* we imagine the possibility of a greater Reality? Then perhaps we may attend to it. If not, I'm quite certain that we'll be *lost* in the external products of our own minds [34].

In this connection, it's interesting to note that Daniel Dennett, this seemingly "hard" algorithmic philosopher, makes a surprisingly "soft" and accommodating turn when it comes not to describing the world, but to us *doing good* (or evil). What he says is curiously consistent with what I'm driving at, on one level at least. At the end of his brilliant book *Darwin's Dangerous Idea,* he takes great care to make the important point that, in practice, *what is good cannot be arrived at by argument.* He calls for "conversation-stoppers": admonitions that will make the infinite "but-saying" of our rational mind *stop dead,* and make us conscious of other human values, ones that we cannot afford to lose. He notes:

> This is a matter of delicate balance, with pitfalls on both sides. On one side, we must avoid the error of thinking that the solution is *more rationality,* more rules, more justifications, for there is no end to that demand. Any policy may be questioned, so, unless we provide for some brute and a-rational termination of the issue, we will design a decision process that

spirals fruitlessly to infinity. On the other side, no mere brute fact about the way we are built is — or should be — entirely beyond the reach of being undone by further reflection (Dennett 1996: 506).

What all this amounts to in relation to us as conscious persons, is that what really matters in our life together is the question: On *what* do we actually base our deepest, most heartfelt hopes and fears? What is the real ground of our lives? When we've become able to intelligently converse on such topics, we may begin to be, not so much unwitting prey to the actual future consequences of our imaginings, but rather the "witting" originators of a world fitted for human beings worthy of the name. This imagined future development must, I think, be based on a sophisticated *rapprochement* between the deepest concerns of modern science (undistorted by the urge to implement for commercial or technocratic purposes) and the deepest insights of traditional religions, of which Eastern Orthodox Christianity has here served as a representative. Dennett's discussion, as I read it, essentially leads to the same position (or, more precisely, to a position consistent with what I'm saying):

> Is this Tree of Life a God one could worship? Pray to? Fear? Probably not. But it *did* make the ivy twine and the sky so blue, so perhaps the song I love [35] tells a truth after all. The Tree of Life is neither perfect nor infinite in space or time, but it is actual, and if it is not Anselm's "Being greater than which nothing can be conceived," it is surely a being that is greater than anything any of us will ever conceive of in detail worthy of its detail. Is something sacred? Yes, I say with Nietzsche. I could not pray to it, but I can stand in affirmation of its magnificence. This world is sacred (Dennett 1996: 520).

Any Christian ascetic would certainly agree: you don't *worship* creation, but it's no less sacred for that, and man (because he is created in the image of God) is cosmically responsible — or even "*pre*sponsible" — for everything he puts his mind to. The curious thing about Dennett's stance is that he obviously feels compelled to invoke the category of the sacred, while denying it any real source other than his personal feelings. This sentimentality, although obscurely indicating a dubious source of "conversation-stoppers," is a far cry from the traditional faith in *God as reality*. The last word, then, appropriately belongs to one raised to this reality, St. Gregory Nazianzen (*d.* 389):

> From the day whereon I renounced the things of the world to consecrate my soul to luminous and heavenly contemplation, when the supreme intelligence carried me hence to set me down far from all that pertains to the flesh, to hide me in the secret places of the heavenly tabernacle; from that day my eyes have been blinded by the light of the Trinity, whose brightness surpasses all that the mind can conceive; for from a throne high exalted the Trinity pours upon all, the ineffable radiance common to the Three. This is the source of all that is here below, separated by time from the things on high [36].

This, according to the ascetic thinkers, is where the sacredness of the natural world, and of human life, really comes from, nowhere else. And if it isn't so, it may well be that the sense of the sacred (and hence of a greater reality) is ultimately nothing but a social construct, precariously based on biologically evolved physiological mechanisms governing intragroup social behavior, and it can be consistently argued that it's best accessed by means of commercially distributed pharmaceutics or other clinical measures [37].

Acknowledgements

This version has benefited considerably from conversations with Jack Cohen, Greg Bear, Kurt Johansson, Pernilla Ouis and Alf Hornborg. The view expressed remains my sole responsibility, of course.

References

AXELROD, Robert. 1990 (1984). *The evolution of cooperation.* London: Penguin Books.

BROOKS, Rodney A. and Lynn Andrea STEIN. 1993. *Building brains for bodies.* A. I. Memo No. 1439, MIT Artificial Intelligence Laboratory.

CASTELLS, Manuel. 1996. *The information age: Economy, society and culture, vol. 1. The rise of the network society.* Oxford: Blackwell Publishers.

CASTI, John L. 1997. *Would-be worlds: How simulation is changing the frontiers of science.* New York: John Wiley and Sons, Inc.

CHALMERS, David J. 1996. *The conscious mind: In search of a fundamental theory.* Oxford: Oxford University Press.

CHANNELL, D.F. 1991. *The vital machine: A study of technology and organic life.* New York: Oxford University Press.

COCKING, J.M. 1991. *Imagination: A study in the history of ideas.* London and New York: Routledge.

COHEN, Jack and Ian STEWART. 1995 (1994). *The collapse of chaos: Discovering simplicity in a complex world.* London: Penguin Books.

– (1997) *Figments of reality: The evolution of the curious mind.* Cambridge: Cambridge University Press.

COLLIANDER, Tito. 1983 (1960). *The way of the ascetics.* London and Oxford: Mowbray.

CROSBY, Alfred W. 1997. *The measure of reality: Quantification and Western society, 1250–1600.* Cambridge: Cambridge University Press.

DAMASIO, Antonio. 1996 (1994). *Descartes' error: Emotion, reason and the human brain.* London: Papermac.

DASGUPTA, Subrata. 1996. *Technology and creativity.* New York and Oxford: Oxford University Press.

DAWKINS, Richard. 1976. *The selfish gene.* Oxford: Oxford University Press.

– 1988. *The blind watchmaker.* London: Penguin.

– 1997 *Climbing Mount Improbable.* London: Penguin.

DENNETT, Daniel C. 1994. The practical requirements for making a conscious robot. *Philosophical Transactions of the Royal Society A,* 349: 133–46.

1996 (1995). – *Darwin's dangerous idea: Evolution and the meanings of life.* London: Penguin.

DIETRICH, Eric, ed. 1994. *Thinking computers and virtual persons: Essays on the intentionality of machines.* San Diego, CA: Academic Press.

EMMECHE, Claus. 1994. *The garden in the machine: The emerging science of artificial life.* Princeton: Princeton University Press.

EPSTEIN, Joshua M. and Robert AXTELL. 1996. *Growing artificial societies: Social science from the bottom up.* Washington, D.C.: Brookings Institution Press; Cambridge, MA: The MIT Press.

GARDNER, Howard. 1985. *Frames of mind: The theory of multiple intelligences.* London: Paladin Books.

HERBERT, Frank. 1976. *Children of Dune.* New York: Berkeley Medallion Books.

HEILBRONER, Robert. 1995. *Visions of the future: The distant past, yesterday, today, tomorrow.* New York and Oxford: Oxford University Press.

HULL, David L. 1988. *Science as a process: An evolutionary account of the social and conceptual development of science.* Chicago and London: The University of Chicago Press.

JOHANSSON, Per-A. 1997. Evolution, ecology, and society: A heuristic ontological framework. *Working Papers in Human Ecology,* 2. Lund University, Human Ecology Division.

KELLER, Evelyn Fox. 1995. *Refiguring life: Metaphors of twentieth-century biology.* New York: Columbia University Press.

KELLY, Kevin. 1995. *Out of control: The new biology of machines.* London: Fourth Estate.

LANGTON, Christopher G. 1996 (1989). Artificial life; in *The philosophy of artificial life,* M. A. BODEN, ed.: 39–94. Oxford: Oxford University Press.

LEVY, Steven. 1993 (1992). *Artificial life: A report from the frontier where computers meet biology.* New York: Vintage Books.

LOSSKY, Vladimir. 1973 (1957). *The mystical theology of the Eastern Church.* Cambridge and London: James Clarke and Co., Ltd.

MAR GREGORIOS, Paulos. 1988 (1980). *Cosmic man; the divine presence: The theology of St. Gregory of Nyssa (ca. 330 to 396 A.D.).* New York: Paragon House.

MARTIN, James. 1996. *Cybercorp: The new business revolution.* New York: Amacom.

McSHEA, Daniel W. 1996. Metazoan complexity and evolution: Is there a trend? *Santa Fe Institute Working Paper* 96–01–002.

MERCHANT, C. 1983 (1981). *The death of nature: Women, ecology and the scientific revolution.* San Francisco: Harper and Row.

MORAVEC, Hans. 1988. *Mind children: The future of robot and human intelligence.* Cambridge, MA: Harvard University Press.

MORRIS, Rosemary. 1995. *Monks and laymen in Byzantium, 843–1118.* Cambridge: Cambridge University Press.

NASR, Seyyed Hossein. 1989. *Knowledge and the sacred.* Albany: State University of New York Press.

The PHILOKALIA: The complete text compiled by St. Nikodimos of the Holy Mountain and St. Makarios of Corinth, Vol. I–II. 1986–1990 (1979–1981). London and Boston: Faber and Faber.

A prayer book for Orthodox Christians. 1987. Boston: The Holy Transfiguration Monastery.

REGIS, Ed. 1992 (1990). *The great mambo chicken and the trans-human condition: Science slightly over the edge.* London: Penguin Books.

RUSE, Michael. 1996. *Monad to man: The concept of progress in evolutionary biology.* Cambridge, MA: Harvard University Press.

SCHMEMANN, Alexander. 1979. *Church, world, mission: Reflections on Orthodoxy in the West.* Crestwood, NY: St. Vladimir's Seminary Press.

SCHUMACHER, E.F. 1978 (1973). *Small is beautiful: A study of economics as if people mattered.* London: Abacus.

SKYRMS, Brian. 1996. *Evolution of the social contract.* Cambridge: Cambridge University Press.

STORK, David G. 1997. *HAL's legacy: 2001's computer as dream and reality.* Cambridge, MA: The MIT Press.

TALBOTT, S.L. 1995. *The future does not compute: Transcending the machines in our midst.* Sebastopol, CA: O'Reilly and Associates.

TIPLER, Frank J. 1994. *The physics of immortality: Modern cosmology, God and the resurrection of the dead.* New York: Doubleday.

TURING, Alan M. 1937. On computable numbers, with an application to the Entscheidungsproblem. *Proceedings of the London Mathematical Society,* Ser. 2, 42: 230–265.

UNSEEN WARFARE: The spiritual combat and path to paradise of Lorenzo Scupoli, edited by Nicodemus of the Holy Mountain and revised by Theophan the Recluse. 1978 (1952). London and Oxford: Mowbrays.

VON NEUMANN, John. 1966. *The theory of self-reproducing automata.* Urbana: University of Illinois Press.

WARE, Kallistos. 1979. *The Orthodox way.* London and Oxford: Mowbray.

– 1989. *The power of the name: The Jesus prayer in Orthodox spirituality.* London: Marshall Pickering

WARWICK, Kevin. 1997. *March of the machines: Why the new race of robots will rule the world.* London: Century.

Endnotes

The reader will notice that the notes of this essay are rather full (and important). Try as I might, I've been unable to avoid this, since much of the ground I tread on is bound to be unfamiliar to many readers, in one way or another. Furthermore, I've made rather extensive use of quotations; this is to give the reader a little more direct glimpses into the thought-worlds confronted below. I relate specifically to two existing traditions (one very new, the other very old), and my own contribution is restricted to drawing conclusions from their confrontation.

[1] Quoted in Cocking 1991: 146.

[2] Herbert 1979: 380; my emphases.

[3] Double-entry bookkeeping has been identified as one crucial innovation powerfully shaping the evolution of modern societal ecology (see Crosby 1997: 199-223).

[4] "The opposite" mentioned in the Frank Herbert quote above can be interpreted (in our case) as a combination of unintended consequences threatening to become overwhelming, in conjunction with a truly severe selective ignorance (read on!).

On the possibility of a machine "takeover" see the recent book by Kevin Warwick (1997), which is more pedestrian, and hence more believable, than Hans Moravec's (1988).

[5] Martin (p. 5) defines "cybercorp" (his own coinage) as follows: "A corporation designed using the principles of cybernetics. A corporation optimized for the age of cyberspace. A cybernetic corporation with senses constantly alert, capable of reacting in

real time to changes in its environment, competition, and cus-
tomer needs, with virtual operations or agile linkages of compe-
tencies in different organizations when necessary. A corporation
designed for fast change, which can learn, evolve, and transform
itself rapidly."

Elsewhere in the book he creates a very strong analogy between cybercorps and organisms and likens the competitive world
that all corporations "live" in to a global, artificial ecosystem.
Current technology can be adequately described as taking on a
life of its own, I agree with that, and it uses human talent to
further *its own interests*, all in the name of greater profit.

[6] Turing (1937). All modern computers act as universal computing machines. By supplying such a machine with a suitable
program, it can be made to mimic *any* computing mechanism
whatsoever. This means that any process or function which can
be reduced to an algorithmic description, becomes possible to
mimic, or implement, on an electronic, digital computer. Also, the
set of solvable algorithmic problems is invariant under changes
in the actual computer model or programming language. Philo-
sophically and scientifically, the major question concerns whether
everything, particularly including the human mind, can be explained in terms of algorithmically operating functions, or not.

[7] For an accessible introductory overview of the techniques and
their meaning, see Casti 1997.

[8] Of course, the scientific *use* of "intermaths" must be separated
from the cosmogonic *story* that will be in question here. The
former will, I'm sure, give rise to very interesting discoveries, but
it has *no necessary* connection with the fervent faith that seems
to lie behind the algorithmic story of the origin of everything. I
hope the reader will keep this in mind as I proceed.

[9] The concept of implementation is summed up like this by
David Chalmers (1996: 317-18): "Computations such as [combina-
torial-state automata] are abstract objects, with a *formal structure*
determined by their states and state transition relations. Physical
systems are concrete objects, with a *causal structure* determined
by their internal states and the causal relations between the states.
Informally, we say that a physical system *implements* a computation when the causal structure of the system mirrors the formal
structure of the computation."

The actual steps and processes involved are in actuality very intricate and much more complicated, however. It's a real question whether there is anything more than a superficial analogy between the actual implementation of algorithms on computers and what, in evolutionary biology, is conceptualized as "the expression of the genotype in the phenotype" (a position which is itself open to criticism; for an illuminating discussion of basic issues see Cohen and Stewart 1995 and 1997). I won't delve into these issues here, but will simply take the Darwinian algorithmic story of life for granted, as fast becoming (I venture) the mainstream heuristic in relevant fields of study today. Furthermore, although in this essay I place the emphasis on evolutionary computation, much of what I'm saying applies to any computer-mediated or -executed process whatsoever, when seen in its larger societal context.

[10] A universal replicator, i.e., a computational system capable of reproducing any system; a kind of logical definition of the "form" of life (von Neumann 1966).

[11] "INTELLECT (*nous*, [Lat. *intellectus*]): the highest faculty in man, through which — provided it is purified — he knows God or the inner essences or principles [see on "logos" below, this note] of created things by means of direct apprehension or spiritual perfection. Unlike the *dianoia* or reason ... , from which it must be carefully distinguished, the intellect does not function by formulating abstract concepts and then arguing on this basis to a conclusion reached through deductive reasoning, but it understands divine truth by means of immediate experience, intuition or 'simple cognition' .... The intellect dwells in the 'depths of the soul'; it constitutes the innermost aspect of the heart .... The intellect is the organ of [*theoria*], the 'eye of the heart' ... " [*Philokalia*: I: 362].

It should be noted that St. Neilos' expression "bodily concerns" includes rational thinking processes.

"LOGOS [Lat. *verbum*]: the Second Person of the Holy Trinity, or the Intellect, Wisdom and Providence of God in whom and through whom all things are created. As the unitary cosmic principle, the Logos contains in Himself the multiple *logoi* (inner principles or inner essences, thoughts of God) in accordance with which all things come into existence at the times and places, and in the forms, appointed for them, each single thing thereby containing in itself the principle of its own development. It is these *logoi*, contained principally in the Logos and manifest in the created forms of the created universe, that constitute the first or lower stage of contemplation" (*Philokalia*, II: 385).

The "lower stage of contemplation" can be said to correspond in certain ways to pure science (cf note 33).

[12] "REASON, mind [*dianoia*, Lat *ratio*]: the discursive, conceptualizing and logical faculty in man, the function of which is to draw conclusions or formulate concepts deriving from data provided either by revelation or spiritual knowledge ... or by sense-observation. The knowledge of the reason is consequently of a lower order than spiritual knowledge and does not imply any direct apprehension or perfection of the inner essences or principles [cf note 11] of created beings, still less of divine truth itself. Indeed, such apprehension or perfection, which is the function of the intellect is beyond the scope of reason" (*Philokalia*, I: 364).

[13] "Man cannot drive away impassioned thoughts unless he watches over his desire and incensive power. He destroys desire through fasting, vigils and sleeping on the ground, and he tames his incensive power through long-suffering, forbearance, forgiveness and acts of compassion. For with these two passions are connected almost all demonic thoughts which lead the intellect [cf note 11] to disaster and perdition. It is impossible to overcome these passions unless we can rise above attachment to food and possessions, to self-esteem and even to our very body ... " (Evagrios the Solitary, d. 399, in *Philokalia*, I: 39)

[14] " ... you should cut off and kill every passionate attachment to things which, although permissible, are not indispensable, as soon as you notice that they weaken the intensity of your will for good, distract attention away from yourself and disorganize the good order you have established in your life" (*Unseen Warfare*, 107).

[15] The *Philokalia* is a collection of texts written between the fourth and fifteenth centuries by spiritual masters of the Orthodox Christian tradition. It was first published in Greek in 1782, then translated into Slavonic and later into Russian; it first began to appear in English in complete form in 1979.

[16] I've chosen Eastern Orthodoxy because it's intuitively and substantially alien to Western (i.e., modern) modes of thought, whether religious or secular. At the same time, paradoxically, it clashes *less* with the now emerging dynamic conception of the universe than previous Western notions of the cosmos. The similarities should not be exaggerated, however, because the spirits animating the ascetic practice of Orthodoxy, on the one hand, and most of current science and technology, on the other, are very different from one another.

A personal disclaimer is also in order. I'm not myself Orthodox, but a (nominal) Lutheran. The immeasurable value of the Orthodox tradition to me lies in its possessing the intellectual riches and depths sorely needed these days by any Christian or past-Christian confronting the spirit of the times. I apologize in advance to the real bearers of Orthodoxy for any unwitting misrepresentation of the tradition that may appear in these pages.

[17] This (among other things) makes the emphasis of the Orthodox ethos different from the Catholic, and there are also institutional differences; in the Orthodox world there are *no* monastic orders (Benedictines, Carmelites and so on), and *no* "central authority" (no Pope!). There's only an adherence to a common tradition, adapted to local languages and customs, but deeply and recognizably similar everywhere.

[18] It should be carefully noted that the conception (or rather non-conception) of God in the Eastern Church is thoroughly "apophatic," which means that God is *absolutely incomprehensible*, but that He can be experienced in His "energies" (see all the works on the Orthodox tradition referred to here).

[19] Cf. section 8, and notes 31 and 33.

[20] In the sense of Damasio 1996, perhaps; cf. also Gardner (1985).

[21] "FAITH [*pistis*, Lat. *fides*]: not only an individual or theoretical belief in the dogmatic truths of Christianity, but an all-embracing relationship, an attitude of love and total trust in God. As such it involves a transformation of man's entire life" (*Philokalia*, I: 360). This means that, if one *does not* have faith in God — the Creator and Origin of all that exists — one's inherent *capacity* for faith is directed to something else, which then effectively becomes one's god. A proper understanding of this

is very relevant to any discussion of man's use of his creative imagination—whether to the "Glory of God" or to exult over himself as he is ("fallen" and "unrepentant", i. e., without having subjected himself to a change of heart, or *"metanoia"*; cf below).

"Faith comes not through pondering but through action. Not words and speculation but experience teaches us what God is. To let in fresh air we have to open a window; to get tanned we must go out into the sunshine. Achieving faith is no different; we never reach a goal by just sitting in comfort and waiting, say the holy Fathers. Let the Prodigal Son be our example. He *arose and came* (Luke 15:20)" (Colliander 1983: 1).

[22] That is everyone who thinks that evolution somehow discredits Christian thought, or traditional religious thought generally.

[23] I won't pursue this question here; let me just mention that they have in common the conviction that the world itself is incessantly changing and that no permanence is physically possible. Thus there's no necessary opposition, when it comes to the world as such, between the classical Christian view of creatures and the current view of general evolution. The difference is one of perspective and emphasis, and of different conceptions of "origin."

St. Gregory of Nyssa (*d.* 395) expresses the dynamic worldview very clearly:

"Who does not know that human nature resembles a stream, from birth to death ever advancing as by an irresistible movement, and that when that movement ceases, then comes also the end of existence. This movement (*kinesis*), however, is no mere displacement from one locality to another (for how can nature go out of itself?). But like the flame on the wick, which appears to remain always the same ... , but in fact is always wholly passing away and never remains the same ... , so much so that anyone touching the flame twice does not both times touch the same flame ... ; the flame is ever new and being renewed, it passes away every moment without remaining the same and is generated anew every moment. So is also the situation with the nature of our bodies" (quoted in Mar Gregorios 1988: 85).

[24] Quoted in Levy 1992: 271.

Daniel Dennett is involved, with Brooks, in the construction of a *humanoid* robot, named Cog, at the MIT Artificial Intelligence

Laboratory (Brooks and Stein 1993, Dennett 1994). The stated aim of the project is "to build an integrated physical humanoid robot including active vision, sound input and output, dextrous manipulation, and the beginnings of language, all controlled by a continuously operating large scale parallel MIMD computer" (Brooks and Stein 1993: 1). This work is supported, in part, by the Advanced Research Projects Agency (ARPA) of the Department of Defense, by Matsushita Corporation, and by the General Electric Faculty for the Future Award; the societally contextual implications of this should not be overlooked when discussing the meaning of the project.

[25] There's a difference between the orthodox Darwinian and the "complexity" varieties of cosmogonic and cosmological algorithmicism, but this is largely an internal scientific dispute of slight consequence to the general theme pursued here. Also, the intellectual cross-fertilization between camps should be obvious to anyone, I think.

[26] "REPENTANCE [*metanoia*, Lat. *poenetentia*]: the Greek signifies primarily a 'change of mind' or 'change of intellect': not only sorrow, contrition or regret, but more positively and fundamentally the conversion or turning of our whole life towards God" *(Philokalia*, I: 364).

[27] ' ... stillness or inward silence is known in Greek as *hesychia*, and he who seeks the prayer of stillness is called a hesychast. *Hesychia* signifies concentration combined with inward tranquillity. ... it denotes in a positive way the openness of the human heart towards God's love" (Ware 1979: 163).

[28] This kind of criticism has in fact been leveled against our emerging computer-based culture by Stephen Talbott (1995), albeit from a different perspective.

[29] A tentative theoretical foundation for this kind of comparison is laid in Johansson 1997.

[30] This view is well captured and expounded in Dasgupta (1996) and Hull (1988).

[31] Doctrinal orthodoxy, which is essential, is best viewed as a necessary *practical* matter, pertaining to the intellectual domain: "Outside the truth kept by the whole Church personal experience would be deprived of all certainty, of all objectivity. It would

be a mingling of truth and falsehood, of reality and of illusion: "mysticism" in the bad sense of the word. On the other hand, the teaching of the Church would have no hold on souls if it did not in some degree express an inner experience of truth, granted in different measure to each one of the faithful" (Lossky 1973: 9).

Whether one accepts the validity of this or not, it should be clear that the dogmas (or mysteries) of the Church must not be put on a par with scientific or philosophical propositions, but are to be approached as revealed symbolic *signposts* indicating the true and proper direction in which to turn one's aspirations, as well as implicitly warning against false or misleading directions. They also have a symbolic cosmological and anthropological meaning, to be sure, but if they're separated from religious aspiration and practice they immediately become dead letters, and fail to illuminate and guide man's intellectual faculty in moral and spiritual matters. Failure to recognize this mars many an argument comparing religious and scientific world-views.

[32] From a prayer in *A prayer book for Orthodox Christians*, p. 7.

[33] This is perhaps a good place to clarify a few basic assumptions implicit in the main text. First, the place of true science in my argument; second, the relationship between Christianity and other religions.

Science is not a religion, but it's certainly a search for and a love of the truth (or it ought to be, to deserve its name). Because of this, the best scientists through the ages (even if not overtly religious) more often than not seem to have had an intuition of what might be called the *truth of reality*, an awareness that reality and hence truth is utterly independent of human transient opinion, and that in itself it's *non*-propositional. This is basically a religious stance; to have felt by one's own deep (re)search the meaning and import of non-subjective, transcendent truth, is to be already halfway to an appreciation of the deeper revelations of other paths to the same spiritual summit.

As for the proper relationship between Christianity and other religions, I agree with the Islamic scholar Seyyed Hossein Nasr, who writes: "The unity of religions is to be found first and foremost in [the] Absolute which is at once Truth and Reality and the origin of all revelations and of all truth. ... Only at the level of the Absolute are the teachings of the religions the same. Below that level there are correspondences of the most profound order but not identity" (Nasr 1989: 293).

In practice this means that in order for one really to become removed from egocentric opinion and inflation, and brought into the light of truth, it's necessary to stick to *one* tradition, normally the one one was born into. Any self-constructed syncretism is bound to be spiritually fatal. This means "that to have lived any religion fully is to have lived all religions and that in fact to realize all that can be realized from the religious point of view man can in practice follow only one religion and one spiritual path which are at the same time for that person *the* religion and *the* path as such" (ibid., 296). The intellectual corollary of this is that to "carry out the study of other religions in depth ... requires a penetration into the depth of one's own being and an interiorizing and penetrating intelligence which is already imbued with the sacred [in one's own native tradition]" (ibid., 282).

Let me add, as a disclaimer, that while I find Nasr's views illuminating in these practical spiritual matters, I haven't much sympathy for his uninformed and somewhat muddled criticism of scientific evolutionary thought (see Nasr 1989: 234–42). Briefly put, he confuses 1) the theoretical concepts of Darwinism and 2) the concept of evolutionary progress (in whatever sense). These ideas are not identical and can surface in different mutually con-tradictory combinations (cf Ruse 1996).

[34] "[There] is but one essential sin, one essential danger: that of idolatry, the ever- present and ever-acting temptation to absolutize and thus to idolize "this world" itself, its passing values, ideas and ideologies" (Schmemann 1979: 83).

[35] "Tell me why the stars do shine,
Tell me why the ivy twines,
Tell me why the sky's so blue.
Then I will tell you just why I love you.

Because God made the stars to shine,
Because God made the ivy twine,
Because God made the sky so blue.
Because God made you, that's why I love you."

Mr Dennett is even kind enough to provide us with the notes to this traditional little song, in an appendix.

[36] Quoted in Lossky 1973: 44.

[37] To be sure, for science and technology enthusiasts of a mystical bent there are quasi-spiritual alternatives, e.g., Carnegie-Mellon University roboticist Hans Moravec's (1988) scenario of human "downloading" in computers, or Tulane University physicist Frank Tipler's (1994) calculations of the universal simulation resurrection of all mankind at the end of time (the Big Crunch). Cf also the "Principia Cybernetica Project" on the WWW, where, in some documents, cybernetics + evolution is elevated to the rank of pseudo-religious metaphysics. For a hilarious and unsettling review of the many visionary links between crackpot science, pseudoscience and real science-to-be (the latter including some of the kind we've met in this essay), see Regis 1992.

Of course, the only valid reason for speaking of *quasi*-spirituality and *pseudo*-religion is a recognition of the real existence of actual spirituality and true religion. In this respect one cannot beat about the bush, and stay honest. When it comes to faith there is a choice to be made, and we all face it. At rock bottom this choice is between different cognitions of reality, if "cognition" is taken so broadly as to include social and hence moral action.

7

EINSTEIN AT THE AMUSEMENT PARK: THE PUBLIC STORY OF RELATIVITY IN SWEDISH CULTURE

KJELL JONSSON

Stories about science can exercise a deep influence on and in turn be influenced by other cultural spheres, even those which at first glance seem very far removed indeed from laboratories and seminar rooms. In the present essay, one particular scientist and his theories will be discussed as a public story of science. At the centre stands Albert Einstein and his theories of relativity, and the examples provided here are primarily taken from a "Swedish story" told during the first few years of the 1920s.[1] The story of one Albert Einstein and something called "relativity" dominated the public image of science in Sweden during the first half of that decade [2].

Year after year, famous physicists and other scientists had urged the Swedish Nobel Prize committee to award the prize in physics to Einstein. Many members of the international scientific community began to express irritation at the niggardly attitude of the Swedes toward the great man. By 1921, the time it seemed had finally come, but due to a lack of unanimity within the committee, it was decided not to award a prize in physics at all that year. The following year, the decision was finally made to award the

Nobel Prize in physics for 1922 to Niels Bohr, and the suspended prize from 1921 to Einstein. However, he did not receive the prize for his theories of relativity, but for his explanation of the so-called photoelectric effect instead, which was something the experimentally minded Swedes could accept as real physics. Still, many felt that Einstein had received the prize too late and for the wrong reason.

Einstein failed to give an acceptance speech at the Nobel Prize ceremonies in Stockholm in December 1922. Rather than confront the icy December weather of the Swedish capital, he chose to embark on a month-long tour of Japan instead. The years between 1919 and 1923 were in general hectic ones for Einstein. He visited Holland, England, the USA, France, Palestine, Spain and of course, Japan. He was on promotional tour, much like a pop artist who had just released a new album. This behavior galled many observers. The amateur physicist Sten Lothigius, who had founded the Swedish Society of Physicists in 1920, composed a pamphlet against Einstein in 1922, wherein he accused the latter's theories of being nothing more than frippery and abracadabra. He accused Einstein of sacrificing scientific truth on the altar of his quest for personal fame, which he had furthermore sought to accelerate through open and covert advertising in the daily press.

In the second week of July 1923, Albert Einstein finally arrived in Sweden.[3] He came to attend the 17th Scandinavian Natural Scientists' Meeting, which coincided with the celebration of the 300th anniversary of the founding of the city of Gothenburg, where it was held. Einstein stayed for four days in the city, during which sojourn he held numerous lectures, including the long-awaited Nobel speech, which was given on July 11 at the local amusement park, Liseberg.

The day the 17th Scandinavian Natural Scientists' Meeting was ceremonially convened was a hot one in Gothenburg. No one was in any hurry to find his seat in the gigantic greenhouse-like conference hall, where the sun beat mercilessly down on the audience through enormous windows, causing one's clothes to stick to the newly-lacquered seats. Fortunately, the first lectures dealt with cool subjects: vitamins and the Ice Age. The meeting boasted 600 active participants dividing their time between eight different sessions. However, among the audience one could also find high-school teachers there to hear their old professors or acquaintances from their student days who had continued along the scientific path to the laboratory and the academy. Women were also present in significant numbers, as the papers duly reported.

Einstein arrived late. He was supposed to have spoken at the inaugural banquet on July 9, but his train from Copenhagen had been delayed and he was forced to discreetly assume his place at the head table alongside the likes of the King of Sweden and the renowned chemist Svante Arrhenius. The day after the inaugural festivities Einstein held a lecture entitled "Zur Weyl-Eddingtonschen Theori" at Chalmers College of Technology. A number of his listeners apparently made comments on his speech, including Arrhenius and the Norwegian physicist Vilhelm Bjerknes. Arvid Hedvall, a professor at Chalmers, stated that he hadn't understood very much of the lecture. A reporter in attendance from the major morning newspaper *Dagens Nyheter* probably hadn't understood much either, choosing instead to simply write about what he saw: "His face could have been the face of a musician or a poet instead of a scientist ... what a voice this man possesses! Soft, mild and caressing. One might have thought that he was reciting a poem by Heine when he in fact was elaborating the most convoluted hypotheses." It was understandable that the reporter should capitulate. Even a writer from *Scientific American* had done so, explaining that when we attempt to understand what it is all about, we eventually give up and are content with musing over the much simpler, though even more unsolveable problem: "How has Professor Einstein succeeded in acquiring and maintaining such a fantastic head of hair?" [4]

July 11, 1923 finally arrived and Albert Einstein would deliver his extremely belated Nobel speech. He did so in the conference hall of the ongoing anniversary exhibition, the so-called "Concert Hall," designed by the Finnish architect Alvar Aalto, later destroyed by fire in 1973. When Einstein began his speech on the topic of the theory of relativity, the entire conference stood still. Zoologists came, physicists, chemists and physiologists from the Chalmers Institute came, astronomers, geographers, botanists and meteorologists came. Most were wearing light-weight summer suits and white dresses, others wore tails. A thousand men and women were in attendance. This time, the newspapers made a more concerted effort at explaining what Einstein's ideas were all about. Einstein discussed the Galilean principle of relativity and the principle of the constancy of the velocity of light. He also discussed the concept "simultaneity" and touched on the general theory of relativity and the question as to whether the universe was closed, finite or infinite, and on the curving of light when it passes through a field of gravitation.

Einstein received a hearty ovation both before and after his lecture, which had taken ninety minutes. "The silence was total," one of the newspapers wrote, "and when Professor Einstein attempted to at least somewhat concretize his lecture, he made elegant gestures upwards toward the hall's capacious vault, when he spoke of various coordinate systems or pointed at the edges or middle of his lecture in order to elucidate his definition of the concept *simultaneity,* many listeners stretched forward in expectation, their comprehension perhaps already overtaxed, as if this were some sort of spiritist experiment where the materialization of the dead was soon at hand."[5]

*   *   *

The public and popular image of Albert Einstein and his most significant contribution to theoretical physics — the theories of relativity — has undergone changes during the 20th century. Up until 1919, the man and his theories were unknown to the public at large. Just one year later, Einstein was world famous. He was portrayed at one and the same time as a revolutionary, a publicity seeker and an impossibly learned scientist, almost a magician. The theories were said to have completely transformed our apprehension of reality. Toward the end of the Second World War, his name was for the first time linked to materially destructive forces and the equation $E = mc^2$ was claimed to lead directly to the atomic bomb. On July 26, 1993, the Swedish evening tabloid *Expressen* featured an article on our century's most famous natural scientist. The headline read: "Albert Einstein: A Male Chauvinist Pig." The article stated that we were wrong to praise Albert Einstein as a scientist of genius. First and foremost, he was an unsympathetic male who made his wives unhappy, was mainly interested in "vulgar women," was a sex maniac, couldn't have cared less about his children and never washed. Nowadays, we live in a culture of mass media, which exhibits a huge interest in the private lives of celebrities and often pays more attention to other things than their public works. Could it be that Einstein will be remembered during the next few years for his sloppy personal hygiene?

When the name and scientific works of Einstein turn up in newpapers, radio and TV today, it is often in connection with occurrences in the field of the natural sciences or in advertising. It is also common that Einstein's name and face personify extreme talent or incredible insight into the mysteries of the cosmos. When we hear the words "Einstein" or "theory of relativity" we begin to associate to the atomic bomb; we see Einstein's great mane of gray hair in advertisments for computers, adult education and Carlsberg beer; or we recall one of our century's most hackneyed slogans: "Everything is relative."

The story of Albert Einstein contains several mythological elements. The myth is based on the idea that the scientist, like the magician, one day suddenly discovers the truth about the world. Roland Barthes has written that the myth of Einstein was intimately connected with the perception that there existed only *one* secret in the world and that secret was to be found in the strongbox that was the universe. Einstein was very close to discovering the key to this safe, the "Open, Sesame!" that would cause the great walls barring entrance to the cave to retreat. Einstein almost found it, and that is what the myth of Einstein is all about; all the necessary Gnostic features are there: the unity of nature, the opportunity for a fundamental reduction of the world, the power of words to open something, the ancient struggle over a secret and a word, the idea that total knowledge can only be revealed once and for all, just like a lock that suddenly springs open after thousands of fruitless attempts [6].

Coverage of Albert Einstein's visit to Sweden was intense. In a way, it was symptomatic that he should end up speaking at an amusement park. Albert Einstein was already a celebrity, even a cult figure, in many countries. His fame was well on its way to being established for all time. In the early 1920's he already belonged to the select few to whom any and all questions could be posed: On the existence of God, the future of the world, on good and evil, the meaning of life and the Swedish disposition. His public image and revolutionary scientific ideas exercised a strong attraction after the First World War. Apparently, a gigantic appetite for new ideas existed at that time. Political thinkers, psychologists, philosophers, writers and artists everywhere formulated subversive thoughts. Communism, fascism, Fordism, dadaism, futurism and psychoanalysis were commodities in high demand in an era seeking fast—but above all new—answers to the riddles of existence. Einstein and his ideas fit into this pattern. The man and his work became an integrated part of the story of cultural modernity. He provided a positive example to many, while others included relativity among the already long list of dreadful aspects of modern life: abstract art, positivistic philosophy, sexual freedom, the dissolution of norms, and bolshevism.

The cultural reception and mythologization of the man and his ideas was influenced and reinforced by journalistic sensationalism.[7] However, once interest arises and stories begin to take form, they can also obtain further nourishment from the opinions of other well-known scientists and philosophers. Nor are Einstein's own pronouncements and actions insignificant in this context. His participation in spectacular events, his hectic traveling schedule, his lecturing and last but not least his bohemian appearance all seem to confirm that this was a remarkable individual indeed, a genius of a scientist who in his theories of relativity had formulated the truth about the universe, life and everything else. Einstein was a public figure who never hesitated to use his fame in the services of a good cause. He gladly exploited his celebrity in order to comment on religion, the world situation, peace and politics in general.

Prior to 1919 very few people outside the scientific community (and not too many within it, either) had heard the name Albert Einstein. Fewer still had any knowledge of the effect of the theories of relativity on the physical image of the world. After 1919 (in Sweden, after 1922 in particular), Einstein's face became as well known as his ideas were misunderstood. However, Einstein had formulated his fundamental theories long before this.[8] In 1905, the seventeenth volume of the scientific journal *Annalen der Physik* was published. One of the most frequent contributors to the journal that year was one Albert Einstein, an obscure engineer working at the patent office in Bern, Switzerland. His daily tasks cannot have been too taxing, since he had time to do research "on the sly" to such an intense degree during office hours. The articles he got accepted for publication dealt with three separate areas of theoretical physics. One article broadened and revised the statistical physics of Ludwig Boltzmann; another treated light as a flow of particles (and included the theory of photoelectric effect, which rendered him the Nobel Prize); a third was a study of electrodynamics in bodies in movement and lay the foundations for relativity theory number one.

This is not the right place to attempt a complete or broad presentaton of Einstein's contributions to science. Instead, just a few words about the theories of relativity. The special theory of relativity concerned measurements. It deals with the laws of physics in those cases where different observers move with constant velocity with respect to one another. In the special theory of relativity Einstein showed that an absolute frame of reference or privileged measuring position did not exist; instead, the measurements themselves constitute the system of reference, and only within this system of reference are time and space constant. The measurement of time and space within two coordinate systems moving uniformly and in a straight line relative to one another will not yield the same measurement outcome. The measuring standard will shrink and time will "go more slowly" in the coordinate system that moves faster. Thus the absolute time and space of Newton became dependent, relative to movement and gravity.

The general theory of relativity, formulated ten years later, explained what happened to measurement through acceleration and under the influence of gravity. Light and energy are, according to this theory, also mass. Masses influence one another; therefore light must also be influenced by, for example, the mass of the sun when it passes near its surface. Thus light from a distant star ought to curve slightly, due to the gravitational pull, when it passes by the sun. It was this consequence of the general theory of relativity that a group of English researchers headed by Arthur Eddington investigated in 1919 and whose positive results, cabled out over the world, won Einstein worldwide fame. Just a few weeks after receiving the news that one single ingredient in his theory of relativity had been verified, Einstein and "relativity" were being discussed the world over. The story could begin to be told, and it hasn't reached its conclusion yet. The cultural "use" of these theories of relativity increased through their lack of contact with "reality." The difficulty of using "common sense" in order to imagine speeds near that of light, and time, space and mass as variable entities, or four-dimensional spaces and curved expanses, eventually charged the theories with enormous fascination for a wider audience.

One often hears that it was the popularization of Einstein's theories that led the ordinary reader to interpret his metaphors and analogies far too literally, and to even believe that the theory of relativity proved that "everything is relative." It is always easy to blame the press. The question is whether the mass media are to be held solely responsible for exaggerations in connection with the integration of the theories of relativity into modern culture [9]. Without intending a complete explanation of the "popularity" of Einstein and his theory of relativity and their public misunderstanding, I would like to briefly indicate two circumstances which are signficant in the context. (There are also unexplainable aspects to popularity similar to the sudden interest in bell-bottom pants.)

Einstein's theories were revolutionary, transforming fundamental concepts which previously had been taken for granted both by Newtonian physics and by "human reason." Another important aspect was that the theories were hard-to-understand innovations, even for professional physicists. This revolutionary and innovative aspect appealed to readers who considered themselves to be revolutionary or at least modern and progressive. Austrian Marxists, Italian anarchists, Swedish Young Socialists and Spanish engineers all received news of this revolution in physics with joy [10]. On the other hand, the more culturally conservative were worried by its subversive consequences. Soon you won't even be able to believe in the multiplication table, some muttered. This is not a matter of construing a simple connection between one's attitude toward the theory of relativity and one's political affiliation. In Sweden of the 1930's, both Einstein's most persistent defender, Ragnar Liljeblad, and his foremost antagonist, Harald Nordensson, were politically conservative [11]. The innovative value of the Einsteinian theories also played a role in this context. After the First World War, "innovation" was seen as something positive in itself by many modern intellectuals. In a time of cultural renewal, the theory of relativity could be compared with dadaism, cubism or psychoanalysis. After Einstein, the cosmos seemed very different, and a new physical worldview, which made a clean break with the old one, had been bestowed upon mankind.

The fact that what Albert Einstein had done was so new, so abstract and so difficult to understand that it could not possibly be seen as part of the cultural tradition of the West was an important part of the story of relativity. This story of "incomprehensibility" was composed by journalists and scientists in tandem. Without the aid of scientists — who emphasized that the theory had nothing whatsoever to do with human experience, who stated that it was revolutionary and innovative and that it was based upon incredibly complicated mathematics — the public story of Albert Einstein and relativity would never have taken the form it did right from the start. In particular, the sophisticated mathematics involved served as a veil behind which scientists who were incapable or unwilling to present Einstein's ideas to a broader audience hid. Inaccessibility or incomprehensibility were naturally aspects of the public perception of scientific ideas. This apprehension of the difficulty in understanding science is, as the French sociologist Pierre Bourdieu pointed out, baked into the social definition of the intellectual professions. The public presentation of Einstein's achievements contributed to reinforcing that tendency [12].

Einstein and his theories of relativity also became an irreplaceable component in the ideology of a new generation of physicists, one of many in a complex story of legitimation. At a time when physics, chemistry and mathematics were being accused by the educated middle classes of being too mechanistic, soul-destroying and technologized, Einstein and relativity offered a counter-argument. The historian of science Paul Forman has claimed that when the activities of scientists enjoy high esteem in the eyes of their surroundings, they can in turn ignore the sympathies and antipathies of the cultural milieu [13]. However, when they lose that esteem, they are prepared to reform their scientific ideology so that it coincides with the attitudes and values of the surrounding culture. Aided by the incomprehensible Einstein and his sensational claims about time, relativity and four-dimensional spaces, scientists could link up with an esoteric and highly-prized way of thinking. There they could rest for a few decades in a marvelous space, far from the crass applications and insinuated connections to factories, cannons and the noise of a machinated universe.

*   *   *

The story of Einstein's fame in Sweden began with the English solar-eclipse expeditions. The results of these observations were presented at the Royal Society and were spread throughout the world in a telegram dispatched from London on November 7, 1919. The general theory of relativity was considered as now having been confirmed. Newspapers in England and America were in a state of excitement for days after the announcement of the results of the expedition [14]. Newspapers in Sweden, however, were less impressed by the contents of the telegram from London, if we are to go by the placing of the item in the newspapers several days later. A news item from the world of science just couldn't compete with "Spartacist Riots" and "The Ravages of the Russian Gang." The articles appearing in *Svenska Dagbladet* and *Dagens Nyheter* were hardly more than small notices, located far from the front page; in one instance, hidden under a large advertisement for "Pellerin's Vegetable Margarine." The tone in *Svenska Dagbladet* was positive: "The greatest discovery in connection with the law of gravity since Newton's day."[15] *Dagens Nyheter* was more restrained, to the point of being critical toward the contents of the telegram from the Royal Society [16]. A few weeks later, the newspaper's tone was considerably more enthusiastic and the formulations began to resemble those to be found in the world press: "Never before in the history of human thought has a discovery been made which will so revolutionize our entire way of thinking, as has Einstein's theory of relativity."[17]

Before Einstein was awarded the Nobel Prize in the autumn of 1922, his achievements were rarely covered in the Swedish daily press. Once in a while a small announcement about some sensational consequence of "relativity" crept in; among the more spectacular was a story in *Aftonbladet* about a new discovery which still today helps the evening tabloids increase their circulation — dieting. An English priest "had proved that the movement which occurs when one dances a waltz, according to Einstein's theory of relativity, makes fat people skinny and skinny people fat."[18] If one followed the rotation of the earth while waltzing, one became thinner, while one became fatter by dancing in the opposite direction.

Articles in the conservative newspaper *Nya Dagligt Allehanda* during the years 1921 to 1923 developed into a minor campaign against Einstein [19]. The editor-in-chief himself, Leon Ljunglund, entered the fray. Ljunglund was frightened by the revolutionary elements in the theories of relativity: "Opponents have called him a scientific Trotsky or Lenin, for he shakes the foundations of our view of nature in a manner in no wise less powerful than the way in which the bolsheviks shook society." Without the slightest touch of irony, Ljunglund wrote about "the Einsteinian bolshevization of our physical worldview" and that the result of that revolution "might imply a collapse for the claims of irrefutable truth, which mathematically-inclined phenomenology has claimed since the days of Kant."[20]

To a man, all the leading Swedish philosophers criticized the theory of relativity. Primarily, it was the classical and absolute concept of time and space that the philosophers were unwilling to abandon; furthermore, they were of the considered opinion that Einstein had mistaken fiction for reality. "Objective and universal fundamental time" was defended. John Landquist, an esteemed literary critic, hoped that the French philosopher Henri Bergson would be able to restore some order to the situation: "These Einsteinian postulates on the extension and reduction of time and space, which have amazed the world and which no one has been able to understand in their entirety, ought perhaps to be considered in the future as a cubistic debauchery of mechanics. Bergson has shown that these paradoxes are optical illusions, mathematical magic tricks completely lacking any contact with reality."[21]

Behind the attacks of the philosophers we can discern a full-scale territorial battle between the new and old queens of science. The theory of relativity was not a purely physical theory; rather, Einstein had shouldered the mantle of philosophy when he had spoken about the concepts of space, time and movement. Even a militant Einstein supporter like Ragnar Liljeblad, the director of ASEA, agreed that physicists had been the major epistemological innovators. In *Teknisk tidskrift* he could even suggest that "in the expert committee given the task of naming a new chair in philosophy, there ought to be a theoretical physicist present." This was an opinion which in turn caused industrialist Harald Nordensson (a close ally of the Uppsala school of philosophy) to cry: "This is an outrageous attack on the Swedish university representatives of an entire discipline."

Nor did all physicists rally around Einstein's ideas immediately or with unabashed enthusiasm [22]. His German colleagues viewed him with scepticism. The generally conservative British physicists were either negative or indifferent. The French, as usual, had their own theoreticians, "hometown relativists" like Henri Poincaré and Paul Langevin. Most Swedish physicists also initially displayed a wait-and-see attitude toward Einstein's theories. When Sweden's most famous natural scientist, Svante Arrhenius (1859–1927), was interviewed, he stated that "the Einsteinian theory is pure speculation, if not to say a piece of theory ... In order to be able to understand four-dimensional spaces, one most likely need be constituted in an altogether different manner."[23]

The status of theoretical physics was weak in Sweden, where the experimental phalanx dominated. For the latter, the theories of relativity appeared to be speculative, or at the very least, highly unsubstantiated hypotheses. Among the more vocal critics of Einstein, a number of senior scientists could be heard, who preferred to formulate their refutations in the conservative press. Before 1919, Einstein's actual supporters were a mere handful. Among them we note in particular Henning Pleijel (1873–1962), professor of electromagnetic theory, and Carl Wilhelm Oseen (1879–1944), professor of mechanics and mathematical physics. Among the small group that later would join them were more youthful celebrities like The Svedberg (1884–1971), Manne Siegbahn (1886–1978), and Oskar Klein (1894–1977).

In a lecture, Oseen indicated that the generation to which one belonged exercised a significant influence in this matter. Aside from Max Planck, he stated, "only young individuals have developed this theory, gripped by its beauty ... Einstein probably perceived theory as the goal of physics, which is so very typical of the young."[24] In the modern discipline of the philosophy of science, scientific change has often been coupled with the thought collective which a generation comprises. The late Thomas S. Kuhn wrote that under normal circumstances, science is a very traditional activity. Thus, fundamental innovations of the calibre of the theory of relativity have a hard time getting a fair hearing. Senior researchers in a scientific culture have a tendency to always solve new problems with the aid of the "old" paradigm.

It is often necessary that a new generation first achieve a position of influence within a discipline or field of research before the old paradigm can be succeeded by a new one.[25] The importance of the generational factor has also been stressed in a very consistent manner by the American sociologist Lewis S. Feuer, in his much-discussed study of the generation of physicists who created "the new physics"—Albert Einstein, Niels Bohr, Werner Heisenberg and Louis Victor de Broglie.[26] All scientific revolutions, according to Feuer, are to a large degree the revolt of youth against "the parental generation."

However important generational belonging and generational experience actually is for understanding the defenders and opponents of the theory of relativity, there are still other aspects which need to be taken into consideration. For example, there are several internal scientific reasons why physicians expressed reservations about the theory of relativity, e.g. that relativity was considered to be a mathematical theory and not a physical one; that the theory had not been verified experimentally to the degree that it could be considered scientifically signficant; and that it challenged fundamental truths about time and space (absolute simultaneity and Euclidean geometry).

Thus, it was the more critical viewpoints which dominated among the Swedish corps of physicists. The "young" Hilding Faxén's (1892–1970) opinions were legion. He considered this theory, "while a thing of beauty from a mathematical viewpoint, to be unhealthy speculation so long as it is theoretically impossible to determine new, provable physical laws or lay a simple foundation for the old ones, neither of which has yet been done."[27] The arguments vary in detail, but the unifying element in the criticism of Einstein's theories is a defense of classical mechanism and the perception of time and space which had been developed within the Newtonian paradigm. The theories of relativity were odd constructions, a threat to common sense and to the comprehensibility of the world.

There has been much speculation about the reasons for the opposition of established Swedish physicists. One much-discussed interpretation was published in the august French journal *La Recherche* in 1983, where cultural factors are particularly emphasized.[28] In this version, German cultural influence on Swedish researchers explains the opposition to Einstein. The

Swedish Nobel Prize committee always favored the Germans, and the thwarting of Einstein is explained by the fact that as a Jew and a pacifist, his name was anathema in right-wing, German nationalist circles, whom the Swedes were unwilling to upset. However, these suppositions are impossible to verify from the available sources, which the historian of science Elisabeth Crawford has proven in a sharply-formulated reply.[29]

The refusal of the Nobel committee to award its prestigious prize to Einstein must instead be seen against the backdrop of national peculiarities within Swedish physicist culture. The committee was dominated by experimental physicists, who found it nigh impossible to accept theoretical novelties.[30] Alvar Gullstrand, given the task of investigating Einstein's contributions to physics, concluded that "faith" determined whether one accepted or criticized the theory of relativity. Naturally, a faith cannot render its originator the Nobel Prize in physics.[31] Even if one refrained from banishing the theories of relativity to the worlds of religion or magic, they were still not considered to be physics in the strictest sense of the word. They were more philosophy than physics. In his speech before actually handing over the Nobel Prize to Einstein, Svante Arrhenius clad this feeling in words: "There is no living physicist whose name has become as widely known as Albert Einstein's. Most bewildering of all is his theory of relativity, which to a large degree deals with epistemological questions and has therefore been much discussed in philosophical circles."[32]

Swedish opposition to Einstein was not particularly political. On the other hand, a full-fledged anti-Einstein campaign featuring definite political overtones was launched in Germany, including certain antisemitic features [33]. However, not even his most aggressive critics in Sweden seem to have been influenced by such considerations. The conservative professor O.E. Westin, who criticized Einstein in numerous articles in both the daily press and in scholarly journals, apologized on behalf of science for the fact that the controversy surrounding the theory of relativity "had degenerated into personal persecution in Germany."[34]

Carl-Olov Stawström, who has examined attitudes among Swedish physicists toward the theory of relativity, concludes that the only physicist in Sweden to have attacked Einstein's person was the amateur physicist Sten Lothigius, mentioned above.[35]

Lothigius was a lawyer by profession, but also the founder (in 1920) and eventually treasurer of the Swedish Society of Physicists, whose chairman was C.W. Oseen. In 1922, Lothigius published a pamphlet featuring rabid criticism of Einstein's theories, which he characterized as "abracadabra" and magic tricks. Einstein was accused of having neglected scientific truth in favor of "personal fame ... which he has furthermore sought to accelerate via a notorious advertising in the daily press, both openly and covertly." The theory of relativity lacks scientific ethics, according to Lothigius: "Einstein has attempted to conceal unallowed shortcuts and horrifying mathematical excesses through the means of cheap tricks."[36]

For Lothigius, the theory of relativity represented all the evils of modernity: superficiality, a lack of perspective, an absence of sense and moderation. Instead, these ideals were to be found in classical physics, with its roots in antiquity. "Some day," he wrote, "a golden age will return, when classical mechanics can once again raise its proud head, calm, worthy, balanced and as aesthetically consummate as the facade of the Parthenon gracing the Acropolis of physics." Several years earlier, Lothigius had attempted in vain to convince C.W. Oseen, a follower of Einstein, of the fundamental misconceptions inherent in the theories of relativity. His way of linking "relativity" to incomprehensible modern art would be an oft-used weapon in the rhetorical arsenal of the critics of relativity: "Einstein is to science what cubism is to art. The ancient Hellenes knew better what timeless value was. How remarkable that these times are so susceptible to phantasies ... for the nonce, the classical idea sleeps."

We have seen that during the early Swedish reception phase of the theories of relativity (the interwar period), there existed internal scientific reasons for adopting a wait-and-see attitude to the news. There was also a built-in, healthy scepticism within the scientific community toward theories which had not yet been proven. On top of all that, the dominant experimental tradition in Sweden acted as a delaying factor. However, in this context we should also take note of interesting cultural contexts which can contribute significantly to understanding the unwillingness of Swedish physicists to embrace the theories of relativity. Though few chose to formulate themselves as drastically as Lothigius, there existed a fear that the "news" would lead to, as oceanographer Hans Pettersson said, "a broadening of the gulf between science and that which one usually calls an educated sense of judgment."[37]

It was not just his opponents who coupled Einstein's name to artistic modernity; the advocates of modernism did so, too. Many made references to the new, four-dimensional geometry. Lawrence Durrell felt that it lent support to new literary and artistic forms. He himself had turned to the natural sciences when attempting to complete his *Alexandria Quartet* in a form based upon the theory of relativity. In Sweden, the composer Viking Dahl dedicated an opera entitled *Sjömansvisa* to Einstein. Dahl was busy at the time creating *Gesamtkunstwerk* in the spirit of Richard Wagner [38]. Dahl was certain that we were on our way to understanding the context of everything. In his pamphlet *Pi väg till ett allkonstverk* he wrote that we were en route to a new Romanticism, which united wireless telephones "with cosmic intelligence and the four-dimensional universe of Einstein, the psychoanalysis of Freud and much, much more." Einstein's so-called "four-dimensional space" played an important role here. In *Dagens Nyheter*, Dahl explained: "Einstein is an international personality of gigantic proportions, who has brought monumental new values not just to the sciences. His theories are of ground-breaking importance to art and music, too ... A fourth dimension is now on its way to be discerned in art and music. So far, we only feel this intuitively, and may be able to discuss it only metaphorically."[39]

*   *   *

Is there any abiding interest in studying the reception and dissemination of a scientific theory? Should we not content ourselves instead with analyzing its ability to make accurate predictions, its veracity and its scientific use? An important reason for being interested in the reception of scientific ideas in various social and cultural contexts is that it unveils something important about social and cultural processes in general, and even provides us with knowledge about the natural sciences as social and cultural phenomena.[40]

Sometimes, the reception of scientific theories and the stories told about them say more about the working of the ideas in question than their actual production. Studying the reception of scientific ideas instead of their origins involves shifting one's gaze from the original and intricate to the public and general. Therewith we notice that while different groups and individuals may use the same terminology—"relativity" for example—they each freight these terms with different meanings and interpretations. The variations among the cultural users of the theory of relativity is related to the distance between everyday language and the mathematical reality with which the theory deals. Abstruseness or even "incomprehensibility" in turn become important elements in the reception of scientific ideas and explain the nearly infinite number of possible interpretations we observe. This is related to the fact that the cultural circulation of scientific ideas is unavoidably affected by moods and interpretations which have no logical connection whatsoever with the theory in question. Thus relativity, for example, can be given entirely personal implications which have nothing at all to do with the consequences of the measurements dealt with by the theory of relativity.

Einstein the man and his scientific work are often used and abused in the cultural context. This has contributed to creating new myths about the activities of science and the uses to which a scientific theory can be put. Young anarchists in the 1920's thought that the theory of relativity proved that the world was multifaceted and self-organizing, which lent support to their own passion for freedom and individualism. Ludvig Nordström— Social Democrat, writer, teacher—felt that Einstein's physics were going in the same direction as his own preaching about worldwide social and cultural unity; the theory liberated mankind from individualism and fragmentization. So even though direct political affiliation is insufficient for explaining one's opinion of Einstein and his ideas, we can still see that the metaphysical and cultural perspectives of the actors played an important role when the public story of relativity took form in Sweden.

Einstein and "relativity" can be used to argue that all measurements are subjective; that truth is relative; that science has abandoned all forms of causality and that "everything" is relative. A scientist and a scientific theory can function as symbolic guidance when modern mankind attempts to orient itself in the world. Albert Einstein and his physical theories belong to the circle of "big ideas" which during certain periods can turn up in all imaginable — and unimaginable — contexts. Clifford Geertz has stated that a "grand idee" like that can be used to explain and defend almost any standpoint, but that the intensity of its cultural use varies over time. After the hectic time that was the early 1920's, expectations were toned down and the phenomenon of relativity was relegated (at least temporarily) to its original field of use — the natural sciences. However, Einstein and relativity seem constantly to crop up as the stuff of which myth and pop culture are made. Even today, when the interest in Einstein's personal hygiene and love-life may just be greater than the interest in his science, the man and his work are far from devoid of cultural use.[41]

Translation by Stephen Fruitman.

References

[1] The material I have examined consists mostly of articles from Swedish daily newspapers such as *Dagens Nyheter, Svenska Dagbladet, Aftonbladet* and *Nya Dagligt Allehanda.*

[2] For an overview of the public image of science at the time, see Kjell Jonsson, "Physics as Culture: Science and Weltanshauung in Inter-War Sweden," in *Center on the Periphery: Historical Aspects of 20th-Century Swedish Physics,* Svante Lindqvist, ed., Canton, MA, Science History Publications, 1993.

[3] Einstein's visit and the reception of the theories of relativity in Sweden is presented in Aant Elzinga, "Einstein i Sverige," in *Tvärsnitt,* 1990:3, 2–13.

[4] Joseph B. Nichols, "You Have One Chance in a Hundred to Understand Einstein," in *Scientific American,* 150 (1934), 73.

[5] *Dagens Nyheter,* 12 July 1923.

[6] See the essay "Einstein's Brain," in Roland Barthes, *Mythologies,* Eng. trans., London, Vintage, 1996.

[7] On the reception of relativity and Einstein's influence on culture, see *The Comparative Reception of Relativity,* Thomas F. Glick, ed., Dordrecht and Boston, Reidel, 1987; Alan J. Friedman and Carol C. Donley, *Einstein as Myth and Muse,* Cambridge, Cambridge University Press, 1985; Abraham Pais, *Subtle is the Lord: The Science and the Life of Albert Einstein,* Oxford and

New York, 1982 and Gerald Holton, "Einstein's Influence on the Culture of Our Time," in *Einstein, History, and Other Passions*, Reading, MA, Perseus Books, 1996.

[8] Albrech Fülsing, *Albert Einstein: Eine biographie*, Frankfurt am Main, Suhrkamp Verlag, 1993.

[9] See Jeffrey Crelinsten, "Einstein, Relativity, and the Press: The Myth of Incomprehensibility," in *The Physics Teacher*, Vol. 18, Feb. 1980.

[10] Barbara J. Reeves, "Einstein Politicized: The Early Reception of Relativity in Italy" and Thomas F. Glick, "Relativity in Spain," in *The Comparative Reception of Relativity*, 189–264.

[11] Carl-Olow Stawström, "Relative Acceptance: The Introduction and Reception of Einstein's Theories in Sweden, 1905–1965," in *Center on the Periphery*.

[12] Pierre Bourdieu, "Intellectual Fields and Creative Projects," in *Knowledge and Control*, Michael F. D. Young, ed., London, 1972.

[13] Paul Forman, "Weimar culture, causality, and quantum theory, 1918-1927: Adaptation by German physicists and mathematicians to a hostile intellectual environment," in *Historical Studies in the Physical Sciences*, 3 (1971).

[14] Abraham Pais, 306ff.

[15] *Svenska Dagbladet*, 8 November 1919.

[16] *Dagens Nyheter*, 8 November 1919.

[17] *Dagens Nyheter*, 11 December 1919.

[18] *Aftonbladet*, 25 August 1921.

[19] See for example *Nya Dagligt Allehanda*, 8 March 1920, 17 August 1922 and 19 November 1923.

[20] *Nya Dagligt Allehanda*, 12 February 1921.

[21] See Suzanne Gieser, "Philosophy and Modern Physics in Sweden," in *Center on the Periphery*, 24–41.

[22] See Stanley Goldberg, *Understanding Relativity: Origin and Impact of a Scientific Revolution*, Boston, Birkhäuser, 1984.

[23] *Nya Dagligt Allehanda*, 25 September 1920.

[24] Quoted from Hilding Faxén, "Relativitetsprinciperna," *Tidskrift for elementär Matematik, Fysik och Kemi*, 2 (1919), 100.

[25] See Thomas S. Kuhn, *The Structure of Scientific Revolutions*, Chicago, 1962.

[26] Lewis S. Feuer, *Einstein and the Generations of Science*, 2d ed., New Brunswick, NJ, Transaction Books, 1982.

[27] Hilding Faxén, 89f.

[28] Girolamo Ramunni, "Prix Nobel," *La recherche: revue mensuelle*, 14 (1983), 1256.

[29] Elisabeth Crawford, "Les prix Nobel et la politique," *La recherche: revue mensuelle*, 15 (1984), 130.

[30] See Robert Marc Friedman, "Nobel Physics Prize in Perspective," *Nature*, 292 (1981), 793–798.

[31] Abraham Pais, 795.

[32] *Les Prix Nobel 1921–1922*, Stockholm, 1923, 58.

[33] See Hubert Goenner, "The Reacton to Relativity Theory I: The Anti-Einstein Campaign in Germany in 1920," *Science in Context*, 6 (1993), 107–133.

[34] O. E. Westin, "Einsteins relativitetsteori: En elementär granskning," *Indlustritidningen Norden*, 49 (1921), 56.

[35] Carl-Olow Stawström.

[36] S. Lothigius, *De Einsteinska Relativitetsteoriernas ovederhäftighet*, Stockholm, 1922, 6.

[37] *Göteborgs Handelstidning*, 20 March 1923.

[38] See *Pi väg till ett allkonstverk: Musikkulturella studier*, 2 (Lund, 1924), 15ff.

[39] *Dagens Nyheter*, 4 February 1923.

[40] See Thomas F. Glick, "Cultural Issues in the Reception of Relativity," in *The Comparative Reception of Relativity*, 381.

[41] Clifford Geertz, "Thick description," in *The Interpretation of Cultures: Selected Essays*, New York, Basic Books, 1973.

8

TELLING SCIENCE

Anders Karlqvist

Resting on Your Laurels?

Popular science writings are not an expression of hard work, they are resting on your laurels. Resting on your laurels is as dangerous as resting when you are walking in the snow—you doze off and die in your sleep.

Those are the words of Ludwig Wittgenstein.[1] But he is far from unique in his opinion. The statement reflects an attitude toward science communication which is quite typical in the scientific community. It is risky business to spend time on popularizing your work. It may even be harmful to your career as a scientist. Success as a writer and popularity with a lay audience are met with suspicion by your colleagues.

Science for the Initiated

The idea that science is an exclusive enterprise, to be understood and appreciated by a limited elite of "initiates," is not new. Throughout history the power of supreme knowledge has been the privilege of a select few. The Pythagoreans in ancient Greece were persecuted. Pythagoras himself had to flee from Samos to southern Italy where he established his famous Brotherhood, who regarded themselves as bearers of special secrets.

The threatening power of scientific thinking was most forcefully demonstrated in the bitter clash between the Church and the early proponents of modern science in the 16th century. Giordano Bruno, philosopher and mystic, was executed in Rome in

1600 for his heliocentric belief and his claim that the universe was infinite, containing many worlds like ours. Galileo Galilei barely escaped the same fate by disavowing his conviction that the Earth was not the center of the universe. These views might have undermined the authority of the Church, and it was important to prevent such dangerous new theories from spreading among the people — and perhaps still more important not to encourage a critical, rational mode of thinking based on observations and facts rather than dogmas.

In Splendid Isolation?

Autonomy and independence of religious and political beliefs have become the hallmark of modern science. To keep science separated from external influences has been a prerequisite for safeguarding the quality control of science. It is easy to point to examples in history where scientific doctrines have been infiltrated and harnessed by political ideologies, sometimes with disastrous consequences for society and ultimately leading scientific development down a blind alley. Lysenko is of course the classic example. The specialists, and only the specialists, are truly in command of scientific methods and concepts — the syntax and vocabulary of scientific language — and hence are the only ones who can undertake critical control of the results and decide what is right or wrong. So the argument goes.

This is, however, a line of reasoning which raises some questions and, as a defense for "science in splendid isolation," it can be and has been challenged. The critique would maintain that science, like all other human activities, is more or less a cultural construct. Some would even insist that it is nothing but a cultural construct. One of the most influential inspirations was given by Paul Feyerabend. In his book *Against Method* [2] he launched the slogan "anything goes." More recently the hoax played by Sokal [3] shows that the internal control and peer review system is by no means infallible.

Since science permeates most phenomena in modern society and relies heavily on public funding, it becomes yet more doubtful to claim that science can stand above scrutiny by society, and that access to the scientific process should be totally separated from democratic control. As is evident from recent controversies about nuclear power, genetic engineering and other areas where social and ethical values are at stake, the dialogue across the borderlines of scientific disciplines is important not only for a democratic society as a whole, but also for the internal vitality of science.

Mathematics – a Safe Haven?

Pure mathematics is usually far removed from the political arena and does not seem to cause any particular social problems. Its issues of right and wrong can be settled by the experts in an unambiguous logical manner. Moreover, as G. H. Hardy [4] observed, the study of mathematics is a perfectly harmless and innocent occupation. In mathematics it is also easy to illustrate with past examples the point of "keeping non-experts out." There exists a rich history of mathematical secrecy.

Cossists [5] was an expression used to identify a circle of mathematicians in 16th-century Italy and France. The Cossists were experts in making calculations, a skill of obvious value for business where it could be applied in accounting and other profitable activities. Their secrets were jealously guarded. When Niccolo Tartaglia in confidence told his friend Girolamo Cardano how to solve cubic equations by an efficient procedure, Cardano had to promise not to reveal it to anyone. In spite of this promise, Cardano published the result ten years later in his major opus *Ars Magna*. It led to a bitter and lifelong feud.

The most notorious of all secret-minded mathematicians was Pierre de Fermat, who lived in the early part of the 17th century. His professional career was spent chiefly as a councillor at the Chamber of Petitions in Toulouse, France, but he was also an amateur mathematician of world class. Fermat was extreme in his refusal to divulge his methods and proofs. He challenged mathematicians all around Europe by announcing new theorems he had proved, while not giving the proofs. These were for other mathematicians to rediscover if they could. Fermat's ultimate contribution, which became the most famous riddle in the whole of mathematical history, was noted in the margin of a book, with the comment: *Cuius rei demonstrationem mirabilem sane detex hanc marginis exiguitas non caparet.* [6]

His proposition can be stated as follows: there exist no integral solutions (x, y, z) to the equation $x^n + y^n = z^n$ for integers n larger than 2. Its "hidden" proof has haunted generations of mathematicians until our day.

Fermat's Theorem Meets its Master

The theorem was finally proved by Andrew Wiles and the result published in May 1995. The story of Wiles' work on Fermat's last theorem has been told in a fascinating book by Simon Singh [7]. Wiles secretively devoted almost ten years to the problem, mobilizing and combining deep and difficult mathematics from a number of (seemingly) unrelated fields. He presented his findings at a seminar in Cambridge in June of 1993. Immediately the mathematical community began to work on the proof, checking each logical step and every minute detail. This was probably the most extensive review process ever to take place. Some two hundred pages of argument were scrutinized. A serious gap was noticed, and another year and a half passed before Wiles figured out how to bridge it. The theorem could be laid to rest with relief.

Fermat's last theorem has given rise to a great deal of popular science writing in spite of the fact that it is a rather marginal (pun unintended!) result in mathematics. Easy to state and historically intriguing, it has also spurred a kind of science communication in reverse. Instead of specialists communicating results to the general public, hobby mathematicians have repeatedly told the specialists how to solve the problem. It has been the "perpetual motion" machine of mathematics — and all of these attempts during three hundred years have turned out to be false. Indeed, Wiles' proof is so complicated and the result so profound that it is unlikely that Fermat did discover a correct proof. Much of the mathematics necessary in the modern proof was not available to Fermat.

Fermat represents an antipode to science communication. You reveal the outcome but nothing more, no why and how. No story is being told. Another example in the same spirit is the semi-amateur mathematician Ramanujan, who was mostly out of touch with his potential colleagues. From a small town in India, he moved to Cambridge (thanks to Hardy) and died at the age of 33 from tuberculosis after suffering the wet British winters (a parallel to Descartes, who could not take the Swedish winter and perished in Stockholm!). Ramanujan's notes are full of amazing formulas, disconnected from mathematical theory, but still shown to be true. Again we get no arguments why and how, no story behind the results.

Mathematicians as Foxes

Fermat and Ramanujan are rare exceptions. No accumulation of knowledge would be possible if scientists were not prepared to share their findings in a more explicit and constructive manner. Yet this does not necessarily mean that pedagogical value is given priority. The typical approach in mathematics is to present your theorems and proofs in such a way that they can be checked logically, but not necessarily understood as to how they were arrived at. Carl Friedrich Gauss, one of the greatest mathematicians of all time, was nicknamed "The Fox." Like a fox with its furry tail, he swept away his trail so as not to show how he had reached his goal. What remains is the strictly logical construction, solid and impressive, to be admired. All the mathematical scaffolding is gone. Lars Ahlfors [8], when asked why he did not explain his own mathematical ideas more explicitly, responded: "A magician never reveals his tricks, does he?"

Even if science is not magic and scientists are not magicians, it is natural that the competition for recognition should encourage a certain secrecy and hide-and-seek in the scientific community. For the individual scientist it is vitally important to be the first to publish. Publishing in refereed journals is what counts, besides being first. To be understood and checked by your peers is very different from being appreciated by a wider audience. In the former case, there is little reason to be entertaining or pedagogical—rather the opposite. All irrelevant ideas must be avoided. The reader should not be tempted to wander off with personal speculations. The reasoning should be unambiguous and the message clear, with no room for conjecture. The writer hopes to guide his peers like the car which meets us when we arrive at a big airport: "Follow me."

No Semantics, Please

The rigid discipline has its price. Emphasis on strict logic, and avoidance of redundant explanations, simply tend to make the writing inaccessible for outsiders. It may be so sterile that it does not even attract the aficionados. And it does not work at all for other, less motivated groups. An obvious example is set theory, which was introduced as an approach to teach children arithmetic. The experiment failed completely. Specialized language should wait until it is needed, says Richard Feynman, and the peculiar language of set theory never is needed.

To free mathematics from superfluous reasoning, to sort out all semantic elements which could distract the mathematical mind, and to establish the foundation of mathematics on a purely syntactic basis — such was the program of David Hilbert and Bertrand Russell. They both tried, in different ways, to reconstruct mathematics in a strict fashion so that paradoxes and contradictions would vanish. In short, they sought a logically consistent mathematical system devoid of intellectual fallacies and human feelings. The project was never fulfilled as planned. The final blow was given by Kurt Gödel in his famous proof that not even arithmetic could be constructed in a logically consistent and complete manner. [9] But there is another lesson to be learnt. Russell's and Whitehead's heroic treatise *Principia Mathematica* is almost impossible to read. It is inaccessible in much the same way as machine code for computers. Strict logical rules and mechanical procedures are not tuned to the human mind. We lose interest and attention if we do not see a meaning and a context. The mind cannot grasp complexity without semantics. That is, there must be a story to tell.

Once Upon a Time ...

All real stories begin with these words. Even if they are not spelled out explicitly, this is how minds are triggered and how we begin on our journey as readers or listeners. Storytelling epitomizes how people communicate ideas, facts, experiences with others. As the above example shows, this basic recipe is frequently ignored or violated in the scientific world — sometimes for more or less rational reasons, but most often because of deeply rooted traditions or lack of interest. Mathematics may not be of immediate concern to the general audience, but it provides some good, transparent instances of how to package a scientific message.

This book is about "science as storytelling." The title tacitly assumes that there is a need for communicating science to people outside the narrow realm of specialists, and for counteracting the reputation of science as an exclusive, inaccessible (and boring) enterprise. I will elaborate on this point a little further, in terms of why and how. One could say that most of the articles in this volume relate to the question of how to make science into good storytelling, either by giving examples or by offering more generic comments and suggestions.

Why Communicate Science, Then?

Einstein's theory of relativity is sometimes said to be so difficult that it was initially understood by only three people in the world (one of them being Albert Einstein). This story is a myth, of course. No theory can survive unless it is understood and appreciated by a considerable portion of the scientific community. There is also a growing awareness that the statement "this is too difficult" is a poor defense. A group at Fermilab has taken an initiative to present its work always in plain English: "If we can't explain what we are doing, then we shouldn't be doing it." [10]

However, our notions of explanation and understanding require some clarification. To understand relativity theory in the sense of Einstein, i.e. grasp the ideas in such a way that you would even be able to continue the work creatively, is a rare feat. This is what we expect of a John von Neumann or an Andrei Kolmogorov, but hardly of an ordinary scientist unless you happen already to be an expert in that very field. Understanding in the sense of being able to follow an argument rigorously and judge its value, though, is what you expect from your peers and referees. It is certainly a less demanding skill to be able to check a result than to create the idea!

As the next level, and here we approach the main focus of our interest, we can talk about understanding in the sense of grasping the essence of the argument, perhaps being able to relate it to other fields of knowledge and to see the implications. This sort of understanding is of crucial importance in interdisciplinary communication. Unlike the first two levels of understanding, it would typically call for non-technical language, or at least a common ground where excessive jargon is avoided. As a final step in this classification of understanding, we might talk about understanding as a kind of appreciation of a scientific theory or idea. Such a general appreciation of science should not be dismissed. Ultimately, this is a form of cultural understanding and acceptance which makes scientific activities viable. It must be based on non-technical information with carefully selected metaphors and simplifying "lies" for the purpose of avoiding misunderstandings, misconceived myths and superstitions.

Let us summarize this simple scheme in a table:

levels of understanding	audience	kind of communication
creative	experts in own subject	technical, visionary
evaluative	experts in own discipline	technical, specialized
interpretative	experts in other fields	general scientific
appreciative	non-experts	metaphorical

A Comparison with Music

It may be illuminating to compare this classification with our thinking about music (while concentrating on classical music, lest the parallel become too far-fetched). We can see the composer as the creative "scientist." It is rarely the case that someone adds to another composer's work (the final part of Lacrimosa in Mozart's *Requiem* completed by Sussmayer is an exception to that rule!). Still, we can think of composers fulfilling and further exploring musical styles pioneered by others. It takes a creative mind to be able to do just that. The evaluative level might be less relevant here, although we can regard academic theorists of music as representative.

The musician who plays and interprets the score does have a definite understanding of what he or she is doing. Ability to sing or play an instrument is quite different from the composer's skills, and may lack some of the insights into the composition which were essential for its creation but not for the performance. It is quite possible to give a fine performance of *Das Wohltemperierte Klavier* without knowing exactly how a fugue is constructed. Finally there is the audience, able to listen to and appreciate music. You can surely be a very sophisticated listener without any knowledge of playing an instrument or of writing musical scores. And after all, the vigor of musical life depends on the people who come to concerts and buy records.

The Rapid Growth of Scientific Knowledge

The accumulation of scientific knowledge in modern society has a price, not only in direct financial terms of demanding considerable funding from society, but also in terms of more and more specialization and fragmentation. The unity of science has perhaps always been an illusory goal, but today this vision seems more distant than ever, despite the efforts by physicists to marry the fundamental forces in nature with a Grand Unified Theory. The lack of overview and comprehensiveness presents a greater threat to understanding and wise management than ever before. The increased sophistication and complexity generated by modern technology and research are not matched by an equally sophisticated grasp of the interrelations of the parts. This is at least to some extent a problem of communication. The fragmentation also has a time dimension. The speed at which new scientific results are produced and presented (and old knowledge is made obsolete and forgotten) poses a tremendous challenge. The Czech president and philosopher Vaclav Havel has captured this tendency in a succinct comment: "The world changes so fast that we are continually forced to generalize about things we have never been specific about."

While scientific results can be kept secret for commercial reasons, it is important for the research community to justify its role and support from society by telling the politicians and voters what it is doing and why. The economic and democratic pressure on scientists has become stronger in the post–Cold War era. Less and less of scientific activity can hide behind the secrecy of a diminishing sector of military- and security-driven research. There is a far more blunt request for demonstrating the relation between resource input and useful output.

The Rapid Growth of Everything Else

This, however, is not a trivial task, even if one decides it is a good thing to do. Much has been said about the blessings of information technology and the virtually unlimited access to information over the Internet. In this accelerating flow of information, science plays a minor role and it is probably losing ground: information grows faster than science. It is somewhat ironic that the Internet, which was originally developed as a tool of communication in the academic world (originating from a military communication network), has been swamped by information which in many cases negates the role of scientific knowledge. Competing structures of thought appear, which address existential issues. Paradoxically, modern Western society building upon technology and scientific knowledge nurtures, at the same time, irrationality and superstition to an increasing extent. Scientific knowledge is being relativized, and it becomes all the more important to bring scientific messages to the public.

To keep science secret or aloof from the general public is a poor strategy. Equally damaging is to claim that science can deliver solutions to all human problems. Such promises are bound to backlash. "Pay us today and we will deliver tomorrow" might be a typical offer from scientists, but it is a tactic which proves ever less viable. This is especially true when costs are skyrocketing and the prospect of useful applications is uncertain and remote. A case in point is research on nuclear fusion, and the hope for unlimited sources of energy through control of fusion reactions. Since nuclear research cannot piggyback on military interests as it used to, this situation is getting more pressing. The old contract between the state and the science community is breaking up. Political support for a super-collider or a space station cannot be taken for granted.

On the Borders

There are areas where science has very little to say, and where the use of a scientific disguise sooner or later will be exposed as "the emperor's new clothes." Although it is impossible to decide permanently where the boundaries of science lie (and how they are changing), it is nevertheless important to recognize that such limits exist. They are most apparent in fields which relate directly to human values and experience. Here we find the main complementary expressions of insights into human existence, with art and religion playing a key role in most societies.

What does this border landscape between science and art, between science and religion, look like? What is the interface between science and superstition? These are intriguing issues worth probing in great depth. It is important both to see what science is not, and to approach a border from each side and explore the productive interaction. The border in question is a fertile ground for creative thinking, not just a line of defense. This has been one of the ideas behind the present book—convening scientists, science journalists, people from the humanities and science fiction writers, all with an interest in transcending their own fields.

The transition from science across this borderline can be expressed in various ways. One which is well illustrated in the volume is science fiction. Hard science fiction is storytelling based on scientific facts, but with the freedom to extend the applications beyond what is technically feasible today, and to fill the gaps in current knowledge with ideas that need not be factually true. Scientists themselves are sometimes driven into areas where the distinction between formula and story, or research and myth, becomes obscure. Cosmology provides many examples. The "anthropic principle," as a manner of reasoning about our existence in the world and our ability to observe the universe, has this dual character.

No doubt the advent of computer technology has vastly influenced our thinking and construction of reality. Virtual reality has become a sophisticated approach to exploring the limits of our senses, and made artificial life and artificial intelligence research a field of practical experimentation, not only a subject for philosophical speculation.

The conclusions so far can be summarized briefly. It is a deep-rooted tradition in science to distrust the popularization of science, and to regard science communication as a dubious distraction. Yet there is a growing need to communicate science, both for internal reasons—to counteract the fragmentation in a mushrooming body of knowledge—and for external reasons: to motivate the support for science in society, a support which is increasingly hard to maintain. Even with good intentions, though, scientists might find themselves at a loss in the tremendous flow of information, where entertainment and other catchier messages dominate the media and data channels. The competition for people's attention is ever more fierce. To appreciate the role of science and its creative potential, a fruitful strategy may be to focus on the limits of science: what science is not, and where it meets other forms of human creativity in religion, art or science fiction.

Explaining and Understanding

Writing good popular science is not easy. Simply translating technical terms into colloquial language is inadequate and, in many instances, might turn out to be misleading. It is difficult to find the appropriate coarseness of description, to establish a dialogue without compromising basic qualities of the scientific work or losing the non-specialist's interest. In this respect, there is an essential distinction between explanation and understanding. The latter demands that we are able to relate a topic to what we are already familiar with, using words which already have a meaning for us. This is why quantum mechanics gets so puzzling. Niels Bohr has advised us to refrain from asking about quantum "reality" and stick to the only things which are accessible to us in the quantum world, i.e. measurements. However, this does not satisfy our common sense and it is not a very reassuring way to help people understand. Physicists can explain the results of quantum mechanics, but to understand is a different matter. Albert Einstein was quite unhappy with this so-called Copenhagen interpretation, and the gap between explanation in a technical sense and an understanding of the "real nature" of quantum physics.

The weirdness of modern physics [11] may seem a huge obstacle in telling science to non-specialists (the physicists themselves are already indoctrinated not to ask the wrong questions, questions which are meaningless according to Bohr). On the other hand, the mind-boggling aspects of physics are a crucial asset in attracting attention, and leave plenty of room for creative speculation — as seen in science fiction literature.

How?

So the question remains: how should science be communicated? What can we learn from other fields, and how can such experiences be used to formulate and reach out with the messages? Let me suggest a possible answer, or at least a direction in which to look. It is not a new answer. One might even say it is the most evident and fashionable answer today. But it does produce significant and appealing food for thought. To communicate science effectively, there must be a story to tell. Once upon a time there was an electron ...

Certainly there are areas which inspire grand stories — the origin of the universe, the evolution of life and so on. These are also stories which are written over and over again and which always have their audience. Other examples are not so obvious. What would a story about Pythagoras' theorem look like? As Ian Stewart demonstrates in this book, it is possible to present mathematics in the form of an interesting story which can be understood and appreciated by laymen. But it can also be made incomprehensible by dressing it up in a technical language that gives no associations with the world we live in.

A Human Activity

"My message is that science is a human activity and that the best way to understand it is to understand the individual human beings who practice it." This statement by Freeman Dyson points to a fundamental aspect of storytelling, the human dimension. It is in our lives and human experiences that we find a frame of reference to which all can relate. Art and literature have exploited this basic fact since the dawn of human history. Although the scientist is not the most familiar figure in world literature, neither is he absent. Many stories have been written about scientists and their lives. All great scientists have had their biographers, and key periods in scientific development have been documented in books like those about "the making of the bomb" [12] or "the double helix" [13]. Today there is a wealth of books on the market dealing with storytelling, mixing the lives of remarkable people with their achievements in science or philosophy [14] and describing influential scientific milieus [15]. Common for this genre is that it highlights the scientific process and the people behind the results, rather than the results themselves. Exit Gauss's fox!

The growing popularity of "science as storytelling" (of which this book is a reflection) can be seen as a response to a demand for intelligible, meaningful and entertaining accounts of scientific work, but also as a way for the scientific community to hold ground in the struggle for attention and support. This approach creates opportunities, as well as temptations and risks. For an artist it is quite natural to mix facts with fiction. Truth takes on a different meaning. Art is a lie which reveals the truth. This is how Pablo Picasso saw it. Likewise, facts are not enough to turn science into interesting storytelling. The missing links are often what make science fascinating. Science would never progress unless people were speculating and jumping to conclusions. And unless people outside the circles of experts are invited to take part in this process, science will remain in the shadow of more exciting activities.

Science on Stage

The human dimension of science leads us to explore other forms of presentation than literature. Of course, the lecture itself is a traditional kind of science performance. Everyone who has experienced a superb lecture knows that this can be as stimulating and moving as a good theater play. Lectures as theater are not an original thought. Demonstrations of various sorts are especially akin to the format of theater, as can be seen in the old "operating theaters" where dissections used to take place. But modern technology has extended scientific shows beyond the lecture halls. Indeed, the (almost) live transmission from Pathfinder's excursion on Mars, and TV pictures from inside our bodies or from the ocean floor, have an attraction of their own. They give us the feeling of being part of the scientific process while it is happening.

The lecture is still a limited form of theater. It is usually a monologue with a few questions thrown in at the end. Yet real science unfolds in a social context, where people interact with each other. To criticize and argue is the essence of scientific life. This makes the dialogue an essential feature in the scientific scene. Hence there is a natural place for drama and for bringing science up on stage. Science as a subject for theater can take diverse shapes. One is the use of theater as a pedagogical tool where professional actors convey a scientific message with techniques of visualization and entertaining tricks—often including humor or focusing on scientific controversies, whose ethical implications are ideal for engaging the audience. Cloning, computerization, nuclear energy, and the greenhouse effect are themes well suited to such exposure. Another method is to let the actors interplay with scientists who represent experts and are interrogated on stage in a dramatic manner.

Science as Dialogue

A different kind of storytelling concerns historical people and develops the drama around their lives. Science is mirrored through the activities of these people. A famous example is Berthold Brecht's drama *Galileo Galilei*. In the hands of Brecht, the life and achievements of Galileo become a source of great theater. We would hardly think of Brecht in terms of science communication, but his play does help to promote an understanding of science in society. The theater provides excellent opportunities to reach people through the dramatic form, and thus attract attention to the human dimension of the scientific process with all its intellectual and emotional components.

However, the theatrical format sets quite specific rules for what works on stage and for what can be achieved in demonstrating a scientific activity. Since everything happens "in real time," there are definite limits on the complication and abstraction of reasoning that can be meaningfully conveyed to an audience without losing the essence of the performance. Concepts which need a blackboard to be understood are not ideal for the theater. Above all, there must be a story which can be cast in dramatic form, and a dynamic element which moves the story from beginning to end.

A further characteristic of drama is conflict. The dialogue where different positions are embodied, arguments "pro and con" are raised, and people in the audience have an opportunity to identify with one or the other view, is fundamental. Ready-made facts are seldom convincing, and simple truths tend to be boring. Nor is it necessary to have many actors on stage to achieve the suspense of a conflict. A dialogue can be in the form of one person arguing with himself or (implicitly at least) with an audience. In the play "Beyond All Certainty" [16] which I wrote together with a colleague of mine, Professor Bo Göranzon from the Royal Institute of Technology in Stockholm, an intellectual conflict is the main focus. The play is about Ludwig Wittgenstein and Alan Turing, and the two opposing views of mathematics — as an invention or a discovery. The opening scene takes place at Cambridge in 1939, at the seminar on the philosophy of mathematics given by Wittgenstein and attended by Turing. Most of the script is based on authentic material. The human drama is conspicuous, but actually the intellectual content seems to have been making a greater impact than the more spectacular ingredients, such as homosexuality and Turing's suicide. My experience is that scientific and philosophical ideas can be brought to life not only in books and articles, but also on stage where the persons behind the ideas and the social context in which the scientific process occurs are portrayed imaginatively.

Actors as Scientists

Several seminars have been held in conjunction with the performances of "Beyond All Certainty," involving the actors and researchers from philosophy and the performing arts and creative writing. The following comments are to some extent based on this material.

The play provided a focus for discussions allowing participants from a range of backgrounds to share a common point of
reference. It was a tremendous challenge for the actors to work
with characters whose ideas were hard or impossible for non-
specialists to grasp. After initial doubt and resistance, they came
not only to accept but to enjoy their roles and perform convincingly. This certainly highlights the importance of storytelling. To
play Wittgenstein and express his ideas on stage does not require
you to comprehend his philosophy. More valuable is to have a
sense of his dramatic impact upon the prevailing state of philosophy. It is quite different from trying to present his ideas through
a narrator.

Drama must throw understandable light on matters which are
difficult to talk about. How do you do that? You have to present
ideas in ways connected with people's experiences. You are not
disguising ideas but exhibiting them obliquely, and mere reference to facts is insufficient. You must go beyond the documentary
details. The actual appearance on stage of Turing and Wittgenstein, even if their physical representation was not intended to
create any simple association with the historical persons, still gave
an embodiment of the philosophical confrontation in the play.
It enabled the audience to experience related theoretical issues,
and to identify the speaker before reflecting on his words. It was
rightly pointed out in our discussions that words are validated
by "the signature through the reference to, or imprint of, a real
body in the utterance."

As I have stressed, the dialogue offers an inherently dramatic
and dynamic form. It allows competing claims to be tested against
each other. It invites participation in the play of unfolding meaning, and is a medium most suited to expressing ideas in motion.
Besides, factual fiction and the fictionalizing of facts show the
real power of art. They yield insight into the imaginative life of
our times.

References

[1] The quotation combines two comments made by Wittgenstein,
dated 1942 and 1939-40, found in G.H. Wright (ed.), *Vermischte
Bemerkungen* (1977).

[2] Paul Feyerabend, *Against Method*. London: Verso, 1975.

[3] The physicist A. Sokal managed to publish a fake article in the refereed journal *Social Text* (Spring 1996) with a baffling blend of quantum mechanics and sociopsychological jargon.

[4] G. H. Hardy, *A Mathematician's Apology*. Cambridge University Press, 1940.

[5] From the Italian *cosa*, i.e. solving equations for the unknown "thing."

[6] Translation: I have a truly marvelous demonstration of the proposition, which this margin is too narrow to contain.

[7] Simon Singh, *Fermat's Last Theorem*. Walker, 1997.

[8] Lars Ahlfors, famous Finnish mathematician (*b.* 1907).

[9] Kurt Gödel, renowned Austrian logician and mathematician (1906-78).

[10] J. Womersley, in "Fermilab," *Science*, April 11, 1997.

[11] For an excellent popular account of the problems of quantum mechanics, see Lindley, D. *Where does the weirdness go?* New York: Basic, 1997.

[12] Richard Rhodes, *Making of the Atomic Bomb*. New York: Simon & Schuster, 1986.

[13] Newton, D. *James Watson and Francis Crick: Discovery of the Double Helix and Beyond*. New York: Facts on File, 1992.

[14] A good example is by R. Monk, *Ludwig Wittgenstein-the duty of genius*. London: Penguin Books, 1990.

[15] Ed Regis, *Who got Einstein's office?* (Addison-Wesley, 1987) tells the history of the Institute for Advanced Study in Princeton.

[16] "Beyond All Certainty" has been produced several times in England, Sweden and Norway by the theater company Great Escape in Norwich. A German version, "Jenseits aller Gewissheit," has been playing at the Kellertheater in Innsbruck. The English text of the play is published in Bo Göranzon (ed.), *Skill, Technology and Enlightenment* (Springer-Verlag, 1995).

9

FRANKENSTEIN'S DAUGHTERS

Paul J. McAuley

Introduction

All human cultures create stories which try to explain the origin of their world. The universal occurrence and vast diversity of myths suggest that making stories is a fundamental human activity.

There is a strand in Western thinking which claims that stories and art are unnecessary, for they are not "real" explanations of the world in the way that science and technology are real explanations. According to this Rationalistic philosophy, hunger for narration is a primitive reflex, while science exists in the pure realm of truth. Indeed, there is a strand of literary theory which insists that narrative fiction is crude, and that instead of telling a story, novels should concentrate on dissecting psychological states. That is, in producing models or hypotheses in a pseudo-scientific fashion which are testable by falsification according to their verisimilitude.

But scientific hypotheses are in themselves a kind of story—or a set of stories—that the scientific culture tells itself. There is the Big Bang story, the Quantum Mechanics story, and so on. And of course, the Creation of Life story, and the Cloning story.

Science Fiction (SF) makes stories not about the present world but about the possibilities that scientific discovery invokes. H. G. Wells, in his 1902 essay "The Discovery of the Future," suggested that it was possible to use fiction to explore histories which did not yet exist—which indeed might never exist. SF

139

uses the future as a place where what is known is systematically changed according to given rules, using a process analogous to scientific argument to support the rationale of those changes.

Two branches of science, information technology and genetics, dominate social and cultural change in the late twentieth century. In the 1980's, cyberpunk reflected the importance of information technology. In the 1990's, SF reflects the increasing dominance of the New Biology, from Paul Di Filippo's biopunk fictions [1] (cyberpunk described ways of positively enhancing the body by mechanical or silicon chip implants; biopunk examines a more fundamental consumerist option, change not just of our bodies but of our cells), to radical reappraisals of the extremes of ecological and evolutionary theory [2].

In this chapter, I will describe and discuss the conflict between rationality and emotion in the fictional treatment of biological engineering, and the way in which these two kinds of story-telling reflect contemporary fears about the manipulation of the unknown country of our bodies.

Dolly and Other Xeroxes

In the past few years we have been treated to a flood of newspaper headlines which appear to spring straight from SF stories. Last year it was Life on Mars; this year it was Dolly the Cloned Sheep.

The procedure by which Dolly was created was straightforward enough, but radical because it had never before been successfully done in a mammal. Dr. Ian Wilmut's team at the Roslin Institute in Edinburgh fused a somatic cell, starved in a special nutrient broth to induce quiescence and remove it from the growth cycle, of an adult sheep with an enucleated oocyte, and transplanted the resulting embryo into a host mother [3].

The news of this astonishing feat was not received with rapture. Far from it. The newspapers in Britain responded with a kind of hysteria (my favorite was from the Daily Mirror— *"Can we now raise the dead?"*) and politicians were quick to follow their example. Coincidentally or not, funding for Wilmut's study was not renewed.

As many newspaper columnists were quick to point out, cloning is the stuff of science fiction. But that does not mean that cloning belongs in the realm of unrealizable fantasy. In fact, it means the reverse, for SF stories are after all about what might be possible (even if sometimes unlikely). SF normalizes the future, so that when a prediction is realized, we are already familiar with it. Many SF fans and writers seemed oddly indifferent to the discovery of evidence suggesting that life may have existed on Mars because in their imaginations they had already been there. They were pleased by it; they were excited by it; but they were not astonished by it.

In fact, the reaction caused by poor innocent Dolly (who apart from her origin is an ordinary Finn Dorest ewe, although perhaps she spends too much time in laboratories) was similar to that caused by horror fiction rather than by SF. That is, a visceral or emotional rather than a rational reaction.

After all, clones are not unnatural, and are not perfect and uncanny duplicates, as anyone who has met a pair of identical twins can attest. Many plants and lower animals use cloning as a natural part of their life cycle — indeed, when I was working as a biologist, I raised a clone of my own, the European strain of the cnidarian species *Hydra viridissima*, or green hydra. At one point I had thousands of genetically identical animals in my laboratory but no one protested because (1) hydra reproduce by asexual budding, forming natural clones, and (2) they are not mammals. But even mammals produce natural clones — for instance, the litters of armadillos are all genetically identical, being derived from multiple splitting of a single embryo, a technique which could be useful in the reproduction of human clones.

Furthermore, we live in a culture in which duplication and multiplication of ideas and images is the norm. We are not horrified by a wall of Coca-Cola cans for instance, not even if next to them sits a wall of own brand supermarket cola cans which strive to mimic the branded cola as closely as copyright law will allow. We are not enraged by Andy Warhol's multiple silk screen images. Nor are we horrified by tribute rock bands which copy those from other eras — and even copy dead performers, such as Jim Morrison or Elvis Presley (at the present rate of proliferation, it is estimated that we will all be Elvis impersonators in about fifty years time).

The fact is that most of the ethical problems raised by cloning have already been explored by SF works. Millionaires raising clones to use in spare part surgery? Try Gene Wolfe's *The Eyeflash Miracles* [4] or Michael Marshall Smith's *Spares* [5]. Or millionaires raising a perfect heir to their fortune — that is, copies of their own selves? Try Gene Wolfe's *The Fifth Head of Cerberus* [6], or my own *Little Ilya, Spider and Box* [7]. Producing a perfect copy of a dead person? Try Ira Levin's *The Boys from Brazil* [8]. Or the problem of identity raised by cloning? Try Ursula Le Guin's *Nine Lives* [9], or Louise McMaster Bujold's *Brothers in Arms* [10], or C.J. Cherryh's *Cyteen* [11].

Many of these fictions have pointed out that ethical fears about cloning are unfounded. For instance, you may be able to produce a physical copy of a person, but not an identical copy, for you cannot duplicate the environment in which that person was raised. If you had been able to clone Einstein, you might have gotten a small man with unruly hair who liked playing the violin, but not a world-changing genius.

And even Dolly herself is not a perfect duplicate. Her genetic material comes from a somatic cell, albeit one which has been despecialized by treatment, and it is possible that her chromosomes contain transcription errors acquired during that cell's normal lifetime. And it is not yet clear whether or not the despecialization treatment affected the chromosomal telomeres, the bits at the ends of chromosomes which shorten with each cell division and hence determine how many times a cell can divide — the Hayflick Limit. We do not know if Dolly's lifespan might be affected by either of these factors, but both are possible obstacles to eternal life by cloning. Also, Dolly's mitochondrial inheritance is different from her progenitor, since her mitochondria (which are normally inherited maternally, and contain extrachromosomal genetic material) would be derived both from the original cell and from the enucleated oocyte.

Even evocations of a scenario of a rigid society populated by various lines of identical workers, as in Aldous Huxley's *Brave New World* [12], ring false when closely examined (and not only because Huxley's alphas and betas and so on were not clones). For instance, producing thousands of identical tailor-made workers is not particularly efficient because you would have to be able to predict that they would be required twenty years after conception — even the old USSR attempted mere five year economic plans. Of course, it may be possible to engineer dumb and docile workers suitable for a variety of simple unpleasant or repetitive tasks, such as hamburger flipping, something I have explored in my novel *Fairyland* [13], but at present these jobs are adequately filled by, for instance, sociology graduates.

However, the considerable exploration of the ethics and practicalities of cloning by SF did not normalize the reaction to Dolly. Instead, Dolly is one manifestation of an ongoing negative reaction to genetic engineering and biomedical science by the public—what one commentator in the UK has termed the "yuk" factor.

The Yuk Factor

The popular reaction to Dolly the Cloned Sheep, in which emotion displaced rationality, mirrors the appropriation by horror fiction of those SF images and tropes associated with biological engineering. What we most fear is involuntary change in our bodies—we like to think that we (that is, our self, or mind) are in control of our flesh casings, but even mild illnesses force us to recognize that we can be helpless passengers. Intervention by biomedical techniques are demonstrations of our lack of control of our basic metabolic functions; genetic engineering threatens our fundamental sense of self, and even of our continuity through our children.

Involuntary change of the body (often through infection, with vampirism as a paradigm) is the stuff of horror fiction. This is exemplified in the movies of David Cronenberg, in which new diseases give rise to new and shocking behavior patterns—as in *Rabid* [14], in which Marilyn Chambers becomes a kind of vampire with a syringe or feeding tube sheathed in her armpit; as in *Shivers* [15], in which the population of an apartment block becomes infected with a parasite spread by sexual congress, and turn into ravening maniacs; as in *Videodrome* [16], with its slogan of "Long Live the New Flesh," in which a TV signal induces a brain tumor and hallucinations so violent they may affect reality, and the hero becomes a kind of living reprogrammable videoplayer, fused with his gun.

In all these cases, as with disease, the victim does not choose to be infected and is horrified by the changes in his or her body. This is in direct contrast with the body changes selected by protagonists in biologically themed SF stories (indeed, Greg

Bear's story, "Sisters" [17], turns upon a little girl who cannot choose to change her body in the ways her playmates happily do).

The dichotomy between use of biological tropes by SF and horror fictions is fundamental, arising from the way in which the fictions are constructed. I suggest that SF stories and horror stories about biological engineering differ because the former derive their scenarios by extrapolation from particular cases while the latter use post hoc rationalizations to justify scenarios derived from reflexive emotional response.

Genre Transformations

It is worth considering two examples of this dichotomy in some detail. The first concerns what may be the first true SF novel, although it has long been claimed by horror writers. The second describes the reification by an SF writer of a device derived from horror fiction.

Example 1: *Frankenstein's Creature vs. Frankenstein's Monster.* Most Americans believe that SF started in 1920's with Hugo Gernsback's *Amazing Stories,* but Europeans know that it began far earlier, in a rented villa by the shore of Lake Geneva, in the summer of 1816. I refer of course to the first novel by a young British writer named Mary Shelley. As pointed out by Brian Aldiss [18], *Frankenstein; or, The New Prometheus* [19], may be the first proper modern SF novel, since it deals with the moral, philosophical and political implications of scientific enquiry and experimentation. Mary Shelley, like all good SF writers, built her story around extrapolations from contemporary scientific investigations — specifically, the experiments of the Italian scientist Galvani, who showed that electricity stimulated contraction of excised frog muscles — that electricity and life are intertwined. In contemporary hospital dramas this trope is very much still current, for scarcely an episode passes without a patient being jolted back to life with electrical paddles. And in the experimental procedure which created Dolly the Cloned Sheep, a minute electric current was used to inject the adult cell nucleus into the enucleated ovum, something that fortunately escaped the notice of most newspaper reporters.

In Mary Shelley's novel, the scientist, Victor Frankenstein, uses electricity to infuse life into his creation, which he believes will be a superior kind of human. His creature is an intelligent tabula rasa, but it chooses to be a monster as a response to the revulsion it encounters, with tragic consequences. However, Frankenstein's creature has been transformed in subsequent movie versions of the story, notably those of James Whale [20], from a natural philosopher alert to the ideas of its author's day to a voiceless lumbering near-robot, inarticulate and menacing. In Mary Shelley's novel, the creature chooses evil; in the movie versions, blasphemous dabbling in creation leads to production of a malign monster that automatically rampages out of control. We do not now think of a creature at all, but of the monster whose image was created by the actor Boris Karloff and the make-up artist Jack Pierce. Through this image, Frankenstein's experiments have come to signify science out of control — hubris clobbered by nemesis, to borrow a pithy phrase from Aldiss.

Example 2: *Posthumanism* versus *The Blob*. Not all SF degrades to horror. In Greg Bear's *Blood Music* [21], a scientist's hubris is indeed clobbered by nemesis, but the novel closes with a redemptive transformation rather than a dying fall. A young, egotistical scientist, Virgil Ulam (perhaps a reference to Stanislaw Ulam, who worked on both the H-Bomb and on early genetic engineering), creates a strain of smart bacteria; when ordered to destroy them, Ulam injects his cultures into himself to sneak them out of the laboratory. His body becomes their universe and they begin to re-engineer him, at first pleasantly (they help him win a girlfriend), then more radically. He loses human form and the bacteria spread from his body to others, transforming all but a few humans who remain as witnesses, and eventually re-engineering local space-time and possibly the entire Universe. All human memory is uploaded into a kind of virtual reality. Nothing, we are assured, is lost. *Blood Music* transforms itself into a transcendental SF novel that echoes Arthur C. Clarke's *Childhood's End* [22], in which humanity passes from corporeal existence into a superhuman group mind.

Yet the main trope of *Blood Music* echoes one which originated in horror fiction. In the novel *The Clone* [23] (Kate Wilhelm and Theodore L. Thomas, 1965) or the movie *The Blob* [24], rampaging monsters reduce humanity (or a portion of the population of Chicago or a small town) into a single protoplasmic mass. In both, the monster devours all living stuff in its path, growing asymptotically before being destroyed. It is the rampage of the monster that is important, not the details of its creation.

Here again is the crucial dichotomy between the way SF and horror fiction deal with biotechnology. In Bear's novel, the act of creation is the template for all that follows, and is described in great verisimilitude. In *The Clone,* a particle of hot dog meat in a sewer is transformed into a ravenous gelid protoplasm by a chance confluence of chemicals and medical waste. In the original 1958 version of *The Blob,* the monster is an alien which arrives in a meteorite; in the 1988 remake it is created by biological warfare.

In horror fiction it is the result, not the process, which is important, and the result of meddling with nature is always the same: a monster on the rampage.

Frankenstein's Daughters

Blood Music was the first important SF work to deal with modern techniques of genetic engineering. Prior to that, much SF was concerned with exploring the use of techniques such as eugenics to accelerate the evolution of humanity. As in much else, this was pioneered by Wells, in novels such as *A Modern Utopia* [25] and *Men Like Gods* [26], but it has proven a fertile field for SF. Supermen, or individuals with enhanced intelligence, have been created by gene transformation (A.E. van Vogt's *Slan* [27]), surgery (Olaf Stapledon's *Sirius* [28] and Daniel Keyes's *Flowers for Algernon* [29]) or viral infection (Thomas Disch's *Camp Concentration* [30]). Interestingly, this theme has also been hijacked by horror fiction. While SF's supermen are either tragic or transformative, Richard Bachman's *Lawnmower Man* [31] (at least in the movie version) becomes a rampaging monster which must be destroyed.

The last radical wave of SF, cyberpunk, described ways of postively enhancing the body through choice, by purchase of mechanical or silicon chip implants which give powers to directly tap into the Internet or computers, or gain a new language or another skill, or gain implanted weapons, such as sheathed scalpel blades (reflecting earlier cyborg stories). Biopunk (as it has been called by one radical practitioner, Paul de Filippo) examines a more fundamental consumerist option; change not just of our bodies but of our cells. It is the most extreme form of plastic surgery, and its consequences have been explored in a raft of fictions, most notably John Varley's "Eight Worlds" [32] sequence, and more recently in Bruce Sterling's *Holy Fire* [33], which describes a society in which biomedical applications can extend healthy life indefinitely, leading to a benign consensual dictatorship—only good citizens get treatment, and the young are disenfranchised by vigorous and experienced post-centurians. Sterling has also written a very perceptive short story, "Our Neural Chernobyl" [34] about the spread of an intelligence plague through populations of wild animals, introducing the idea of gene hackers, the biopunk equivalent of cyberpunk's computer hackers.

All of the above explore essentially somatic rather than genetic changes, but SF has also examined the possibility of using genetic engineering to produce new kinds of humanity capable of living in environments hostile to the base form. This was first postulated by James Blish in a series of stories collected as *The Seedling Stars* [35]. He even invented at useful term for it: pantropy. It is now a given in SF that humans may need some genetic tweaking to adapt to new environments: recent examples include Lois McMaster Bujold's micro-gravity adapted humans in *Falling Free* [36], and Stephen Baxter's neutron star colonists in *Flux* [37].

All of these fictions have at their hearts a single question: what makes us human? All are generously inclusive in their definitions.

Conclusion

Horror fiction is primarily about loss of control. Meddling with nature leads to monsters—Frankenstein's monster in the James Whale movies, or hordes of pollution-maddened frogs, or of giant radioactive ants or rabbits. In horror fiction what is important is the consequences of an act of meddling, not the method. Violation of the norm leads to disaster.

SF is about the increase in possibilities through use of science and technology. SF stories are important not only because they explore the ramifications of science; they also familiarize ourselves with new conceptual territory. SF stitches new ideas into the fabric of society, whether utopian or dystopian, whether for the benefit or harm of humanity. In horror fictions we know that something bad will always happen. In SF, anything may happen. Something terrible, perhaps, but perhaps something wonderful.

What is most challenging to contemporary SF writers is that new scientific predictions are being actualized at an ever increasing rate. To borrow terms from software engineering, the gap between vaporware and hardware is decreasing. The moral problems generated by technological and scientific change that SF deals with—the territory it maps—has been subsumed by the present. We have almost used up the host of scientific discoveries predicted at the beginning of the century, and the pace of scientific advance is now so fast that SF has yet to generate sets of predictions about many of the most recent discoveries. We now gaze at a future without preconceptions. This may be good in that ethical debates may not be overshadowed by distorted popular images, as the debate about Dolly the Sheep was dominated by the image of Boris Karloff with a flat head and bolts through his neck. But it means that for the first time this century we no longer have detailed maps of the future.

That there is less time for SF to explore and normalize scientific possibilities is an indication that the advance of science is reaching the point where it is faster than society can absorb. When that happens, it is possible that for the first time science may come under direct control of politicians. It may become exclusively goal-directed. If only to avoid that, science needs SF stories, and SF writers need to look forward, not back.

References

[1] Di Filippo coined the term biopunk, but unlike the godfather of cyberpunk, Bruce Sterling, does not issue regular bulletins or manifestos. The short stories collected in *Ribofunk* (1996) colorfully and often playfully mix tropes and images from the wilder fringes of popular culture with radical extrapolations of societies based on biotechnology.

[2] For instance, both Greg Bear (in *Legacy* (1996)) and Brian Stableford (in *Serpent's Blood* (1995), *Salamander's Fire* (1996) and *Chimera's Cradle* (1997), which comprise the Genesys trilogy) present ecosystems based on different versions of Lamarkian evolution.

[3] Wilmut, I., A.E. Schnieke, J. McWhir & K.H.S. Campbell (1997).

"Viable offspring derived from fetal and adult mammalian cells." *Nature* 385, 810-813.

[4] Collected in Gene Wolfe, *The Island of Doctor Death and Other Stories* (1989).

[5] Michael Marshall Smith, *Spares* (1996).

[6] Gene Wolfe, *The Fifth Head of Cerebrus* (1972).

[7] Collected in Paul J. McAuley, *The King of the Hill and Other Stories* (1991).

[8] Ira Levin, *The Boys from Brazil* (1976); the movie version (1978) kept the same ending in which a deliberate action of Hitler's clone overturns the trope of genetic determinism.

[9] Collected in Ursula K. Le Guin, *The Wind's Twelve Quarters* (1975).

[10] Lois McMaster Bujold, *Brothers in Arms* (1989).

[11] C.J. Cherryh, *Cyteen* (1988).

[12] Aldous Huxley, *Brave New World* (1932).

[13] Paul J. McAuley, *Fairyland* (1996).

[14] David Cronenberg, *Rabid* (1976).

[15] David Cronenberg, *Shivers* (original title: *The Parasite Murders*) (1974).

[16] David Cronenberg, *Videodrome* (1982).

[17] Collected in Greg Bear, *Tangents* (1989).

[18] Brian W. Aldiss, *Billion Year Spree* (1973), later expanded as Brian W. Aldiss & David Wingrove, *Trillion Year Spree* (1986). His argument frames the action of a novel, *Frankenstein Unbound* (1973).

[19] Mary Shelley, *Frankenstein; or, The New Prometheus* (1818).

[20] James Whale, *Frankenstein* (1931) and *The Bride of Frankenstein* (1935).

[21] Greg Bear, *Blood Music* (1985).

[22] Arthur C. Clarke, *Childhood's End* (1953).

[23] Kate Wilhelm and Theodore L. Thomas, *The Clone* (1965).

[24] Irvin S. Yeaworth, Jr., *The Blob* (1958). The original, mainly remembered because of the debut of Steve McQueen, suffered

an unremarkable sequel, *Son of Blob* (dir. Larry Hagman (1971)). The SFX remake, directed by Chuck Russell, was made in 1988.

[25] H.G. Wells, *A Modern Utopia* (1905).

[26] H.G. Wells, *Men Like Gods* (1923).

[27] A.E. van Vogt, *Slan* (first published as serial in *Astounding Science Fiction* (1940); as a novel in 1946).

[28] Olaf Stapledon, *Sirius: A Fantasy of Love and Discord* (1944).

[29] Daniel Keyes, "Flowers for Algernon" first published as a short story in *The Magazine of Fantasy and Science Fiction* (1959), later expanded into a novel (1966).

[30] Thomas M. Disch, *Camp Concentration* (1968).

[31] Richard Bachman, alias Stephen King, disowned the movie version (dir. Brett Leonard (1992)), which has little in common with the original short story "The Lawnmower Man" (1975).

[32] The important works in John Varley's "Eight Worlds" sequence are the novels *The Ophiuchi Hotline* (1977) and *The Steel Beach* (1992), and the stories assembled in the collections *The Persistence of Vision* (1978) and *The Barbie Murders* (1980).

[33] Bruce Sterling, *Holy Fire* (1996).

[34] Collected in Bruce Sterling, *Global Head* (1992).

[35] James Blish, *The Seedling Stars* (1957).

[36] Lois McMaster Bujold, *Falling Free* (1988).

[37] Stephen Baxter, *Flux* (1993).

10

MISSION TO ABISKO

LARRY NIVEN

During the Book Festival at UCLA in April, Greg Benford invited me to Sweden. Abisko is a biological research station in Sweden, 200 km north of the Arctic Circle. I had never heard of it. It's the site of an annual seminar on systems and society. About 25 people, all the place will hold, gather to lecture each other on a chosen topic. Of course all attendees must be approved. Greg Benford had filled his roster, very late, and then three of them dropped out. If I was to fill one of those slots, the organizers of the meeting would have to approve me quickly. Even so, there was some delay. With a few days' slack I could have saved the sponsors some money, buying "economy" tickets on American Airlines and using upgrades to put me with my traveling companions, two science fiction writers named Greg. Getting the chosen topic was like pulling teeth. When I left home I had a generic title, "The Science of Fiction and the Fiction of Science." I interpreted that somewhat loosely, as the topic seemed a little loose too. Mats Forsman, the meeting coordinator, was having trouble with his electronic mail, so I never got an exact description of the topic until after I returned home. Then I learned why all the other attendees wanted to insert the word "lies" somewhere in their presentations. At the time I felt like the only Red who never got the word: "Hang a portrait of Lenin!"

Gregory Benford told me to take a computer. I did. He and I boarded the flight at LAX. The flight stopped in Chicago and picked up Greg Bear. Then it was on to Stockholm. Jack Cohen of England was much surprised to see me among the attendees lining up to fly to Kiruna. I had never been to Sweden, never been north of the Arctic Circle, and had never seen the midnight sun. We arrived on a bright, sunny Sunday afternoon. The sessions ran Monday through Thursday. Abisko is a biological research station 60 km west of the town of Kiruna. There's an active railway line, and what it hauls is mostly iron ore. There are mountains on one side, and a frozen lake backed by mountains on the other. The mountains are covered with ice. The sun went behind mountains at night. Some of us stayed up one night talking until sunrise, at forty minutes past midnight.

The Abisko folk are wonderful hosts. We ate five to six meals a day: the usual three, plus coffee breaks. There's a sauna. It's been a long time since I was at university, and so I found the housing disconcerting. One hotel-style sliver of soap, and one towel; that's it. Shower and toilet are down the hall. Everyone else took it in stride, and I got used to it quickly. But the guy in the room next to mine complained that I'd robbed him of sleep. I led him to believe that a chair collapsing under me caused most of the noise that first night. (The chair was replaced: his doing, I think.) I will now admit that I kept forgetting to take one thing or another every time I went to shower or use the toilet, and the slamming of my door must have been noisy. Double that because I kept trying to lock my door. I just couldn't figure out how to do it. I'm told one could get e-mail in the station. I never developed enough interest: there was too much going on. In Stockholm, later, I tried to use the hotel facilities, and could not.

For my presentation, I attempted to show how to teach science within fiction, and particularly, how to teach a reader how to use his mind for fun. Do this right, and some readers spend the rest of their lives making up their own homework. They will learn. As a demonstration of skilled daydreaming, I offered an explanation of the "missing mass" as presently understood in astrophysics. Another attendee, Phil Baringer, University of Kansas, has now told me why it won't work; and an article in *Science* gives reason to think that the missing mass has been accounted for. So much for my Nobel Prize. But as a high-flying joke it still works.

All speeches were to be followed by questions and comments. During these periods I seem to have developed a reputation for "bullets," that is, for the epigrams and slogans that speakers might project onto a screen so that an audience will remember something ten to fifteen minutes after his speech. What follows is reaction to general discussion:

- I liked Greg Benford's presentation, "The Biological Century." It may be years before I know if I have anything to add. I write slowly. One presenter is disturbed at how we teach science: we teach that the great discoverers went straight to their goals. This could be very discouraging to any recent grad student. We should show the wrong paths they took too. I listened to him say this, and wondered if we would be graduating students in their fifties! There are too many paths to show. Better to include an Introduction in every textbook: *This book cannot show all of the wrong paths and mistakes science has taken. We expect to teach you enough wrong answers by accident.*

- John Casti fears that we teach that answers come as certainties, that we seem to be pursuing a final truth. We should show more of ambiguity. I disagree. "Gravity is pretty dependable"? Of course, we're pursuing a final truth. Nobody who doesn't do that ever finds anything! C.P. Snow speaks of "The Two Cultures" — scientists and humanists — who can never understand each other's disciplines. I think he's wrong. The "hard" scientists are very likely to be as well read as their colleagues in the humanities. What Snow didn't want to remember, and what no humanities professor is likely to remind another, is that they quit physics and chemistry because the work is too hard! It's worth remembering that for most of humanity, science is hard work. You know you're playing a vast game; you're being paid to do it, and that sometimes feels ridiculous. One day you'll be caught. But most humans would run away if you tried to make them work this hard. In particular, it does little good to tell newspersons that they could sell more science and get rich thereby. You could even make them believe it. But they aren't smart enough. They keep getting caught in mistakes, and they hate that, so it's not what they're selling. We need to teach better. Another "bullet": *Not responsible for advice not taken.* Say the truth as best you can, and hope someone is listening. Liars are not your fault. Gullible fools are not your fault. Think of it as evolution in action.

I did a little dozing in one or another afternoon seminar, and once had to stagger away and sleep. They get up early in Abisko: a 9-hour difference becomes 12 hours difference for a late-rising writer of fiction. It never gets darker than a good sunset. Jet lag didn't hurt me much, considering. The conversations were wonderful throughout. I liked the coffee breaks as much as the meals. I deeply appreciate the fact that all the Swedes spoke English. (I gather they tried bilingual seminars and gave them up as hopeless. Too few outsiders speak Swedish.) I believe that every Swedish participant made some opportunity to explain to me that he didn't read science fiction, and there isn't any science fiction at all in Sweden. Traveling with Greg Benford is great. He makes you exercise. We hiked every day along Abisko's one road. I remember observing a great gap in that ring of mountains for some time, before someone used the word "glacier." On Wednesday the Overlords declared a break. Dr. Benford and others went off to ski, but several of us took the train to Narvik in Norway to play tourist. We went through the World War Two museum, and found a story somewhat different from any we'd been taught. We found a high tea: delightful.

Thursday evening we were ferried by car and van thirty kilometers west to a tiny ski resort. There we were shown photographs of the tundra taken through the proprietor's lifetime. It's a great show. The restaurant was wonderful too. Driving home, the van ran out of gas. I was dressed for the Arctic: I wouldn't freeze. Nobody seemed excited. One of us had a cellphone. As there was no way I could fix this myself, I just walked around a little and listened to conversation: "After all, that's our lake. We can just walk home." "Well, but it's thirty kilometers long!" "Oh." "Just a minute. I don't think that is our lake." "I saw somebody run naked across the road back there." I presently dozed in the back of the van. Someone shook me awake and made me get in a car that took a few of us back to Abisko. One of the station people walked to where some locals were using a sauna (and running naked across the road to a snowdrift) and got more transportation.

The two science fiction writers and I spent two nights in Stockholm, courtesy (again) of the meeting organizers. We stayed at the Reisen Hotel, which is in Gamla Stan, the Old Town of Stockholm. My room was tiny. The hotel is elegant and has a perfect sauna, and the location is wonderful. At Abisko I'd been told repeatedly that there wasn't any science fiction in Sweden. In Stockholm we found a science fiction and fantasy specialty bookstore. I haven't seen anything this big or this good since the late, lamented *Change of Hobbit* in Santa Monica. We went there with John Casti—four science-fiction and science-fact writers suddenly descending on them—and signed some books. I went away with some memorabilia for the Los Angeles Science Fantasy Society, and two new Terry Pratchett novels.

Afterwards, John pointed out a restaurant, *Den Guldene Flygen,* and we ate there that night and found it very good. Agneta Lundstrom was at Abisko. She's Director of the Sjohistoriska Museet in Stockholm. Greg Benford struck up a conversation with her there, and so she took us through her museum in Stockholm. Wow! I'm going on memory here. The museum is all built around a Swedish king's warship, the *Vasa,* that sank during the 17th century. The thing was top heavy with wonderful—but massive—artwork and cannons, and it rolled over in the first breeze. They got those expensive cannons up after not very long, but the ship itself wasn't recovered until the 1970s. The floors of the museum surround the ship itself. It's too fragile to be messed with by tourists, but rooms on the original ship have been reproduced and populated with wonderful wooden statues, as if a full naval crew had been turned to wood by some anonymous Medusa variation who leaves no witnesses.

Agneta then took us to her home for lunch. She has a wonderful apartment that is almost a museum by itself. Jack Cohen had mentioned finding a wonderful "rijstafel" restaurant. It's in Old Town; we found it in the phone book. There were no plugs to fit my computer at the hotel. Greg Benford carries a case of plugs, and that got me a battery charge, but electronic mail was hopeless. On our last morning I woke up at five. What the heck, I was looking at a twelve-hour time change anyway. I got a sauna; walked around; eventually tried the library. The restaurant wouldn't open until nine, but there was coffee in the library, as the desk clerk had said. I saw basic generic breakfasts laid out, decided they were for people up before nine and hungry. I ate one. Mistake: a woman appeared and told me that she and her husband had a plane before nine, so they'd ordered these. I went into panic mode. She reassured me: bad idea anyway, she wasn't hungry and it emerged that I was Larry Niven and her husband was a fan. It was one strange trip, and I wouldn't give any of it up for anything at all.

11

SECRET NARRATIVES OF MATHEMATICS

IAN STEWART

Introduction

A few years ago, I came across a paper by Lamport [6], which advocated *structured proof* in the teaching of undergraduate mathematics. A structured proof is written in a formal style, like a computer program. Lamport argues that such a presentation should make it easier for students to verify the logic of a proof.

Some major innovations in mathematics teaching have been unmitigated disasters, such as "new math," which introduced set theory into schools at the expense of more useful skills like algebra and geometry. The mistake was to confuse logic with "psychologic" to assume that mathematical concepts should be constructed in the students' minds in order of logical precedence. The advocates of new math bought, wholesale, the idea that at root mathematics is an abstract game played with symbols according to meaningless rules. It was a bit like teaching music by starting with the theory of harmony instead of singing songs, and it lost sight of all the aspects of mathematics that give it meaning and relate it to experience. The philosophy was reductionist rather than contextual: focused on form rather than meaning, on syntax rather than semantics.

I felt that Lamport's proposal was in danger of falling into the same trap. Its motivation seemed to be to teach students to dissect a proof into tiny component assertions, verifying them one by one. This is fine if you want students to convince themselves that a proof is logically valid, but it doesn't give them an overall feel for the flow of ideas, so that they can devise their own proofs. It may not even leave them with the feeling that the theorem is true. In a musical analogy, it will teach them to verify that Mozart's harmonies follow the traditional rules, or spot when they don't, but it won't teach them to write their own sonatas or symphonies. In the terminology of Cohen and Stewart [3], true understanding involves "complicity" between syntax and semantics, a self-referential process in which each repeatedly responds to and modifies the other, to create an emergent result.

Kelley's Mistake

That said, Lamport makes a good case: for example, the following striking anecdote.

> Some twenty years ago, I decided to write a proof of the Schroeder-Bernstein theorem for an introductory mathematics class. The simplest proof I could find was in Kelley's classic general topology text [5] (page 28). Since Kelley was writing for a more sophisticated audience, I had to add a great deal of explanation to his half-page proof. I had written five pages when I realized that Kelley's proof was wrong. Recently, I wanted to illustrate a lecture on my proof style with a convincing incorrect proof, so I turned to Kelley. I could find nothing wrong with his proof; it seemed obviously correct! Reading and rereading the proof convinced me that either my memory had failed, or else I was very stupid twenty years ago. Still, Kelley's proof was short, and would serve as a nice example, so I started rewriting it as a structured proof. Within minutes, I rediscovered the error.

The Schroeder-Bernstein Theorem is a key result in Cantor's theory of infinite cardinals. It implies that if each of two cardinals is less than or equal to the other, then they are both the same. The story-line of Kelley's proof goes back to the classic text of Birkhoff and MacLane [1].

It starts with two sets, each of which maps one-to-one inside the other, and constructs a one-to-one mapping of one of them onto the other. It does so by splitting each of them into three pieces, and setting up suitable mappings between these pieces.

For details see Stewart and Tall [9] chapter 12. This threefold structure falls out naturally from the idea of an ancestor—specially created for this proof and not, to my knowledge, ever used again anywhere in mathematics. The ancestry of an element in either set is the stuff you get by working backwards through the two mappings, and the threefold partition corresponds to three possibilities:

1. The chain of ancestors stops at some element of one set.
2. The chain of ancestors stops at some element of the other set.
3. The chain of ancestors continues forever.

Not a gripping yarn, I admit, but if you understand the notation and background concepts then it's clear, memorable, and convincing. Any competent technician can fill in the missing steps of the proof to make it logically watertight.

Lamport's article doesn't say what Kelley's error is, and attempts to find out by e-mail elicited the less than helpful response "write out a structured proof and it'll be obvious." Even though I knew there had to be a mistake, it took me several attempts to find it. In the end I found the error without resorting to a structured proof as such, but my step-by-step checks were pretty much what you'd have to write down in a proper structured proof, so I concede the point.

What was the error? Kelley's proof is what is known in the trade as a "sketch," requiring the reader to fill in routine details. In place of the threefold classification just listed, it uses this one:

1. The chain of ancestors contains a finite even number of elements.
2. The chain of ancestors contains a finite odd number of elements.
3. The chain of ancestors contains an infinite number of elements.

This change is so minor that it tends to go unnoticed by those of us who know the story, but it has a major effect on the proof. In fact, it wrecks it. The reason is that the proof depends not on the number of ancestors, but on whether the ancestry process stops or not. If the chain of ancestors goes $x\,y\,x\,y\,x\,y\ldots$ forever, then the number of ancestors is 2 — finite and even — but the process doesn't stop. This muddles up the threefold split, and the proof falls to bits.

There are at least two ways to intepret this tale. One — Lamport's — is that Kelley should have been more careful to check that his sketch was logically correct, and that he couldn't have failed to observe his error if he'd used a structured proof. Fair enough, and point taken. But there's another interpretation, not contradictory but complementary, which is that *Kelley told a good story badly*. It's rather as if he'd introduced the Three Musketeers as Pooh, Piglet, and Eeyore. Some parts of the story would have made sense — their inseparable companionship, for instance — but others, such as the incessant swordplay, would not have done.

So Kelley had the right overall story-line, but fell down on a crucial detail of the plot. He probably thought that his slight recasting of the tale was a harmless simplification, and didn't spend any time thinking the consequences through — always a dangerous practice.

Structured proofs would help a student spot that error — but I don't think they would have helped him or her to produce their own story. In fact, the less well you know the story, the more likely you are to notice that Kelley's version goes adrift. This is why trained mathematicians have trouble spotting what's wrong with Kelley's story: they already know the plot. It's a bit like proofreading your own book for the fifth time: you see what ought to be there, not the typo that is actually there. Structured proofs would have revealed the error, just as a spell-checker would reveal most (but not all) typos. But there's a big difference between being able to write literature — or even passable journalism — and being able to use a spell-checker; and I think there's a comparable difference between understanding a proof and being able to follow a structured version of it step by step.

Math Boffin in Theorem Shock Horror

Students find mathematical proofs difficult. That's hardly a surprise. On the surface, mathematical writing more closely resembles ancient Egyptian hieroglyphics than literature. Symbolic expressions dominate the page, separated by stereotyped phrases hypotenuses, but how you wrote the idea down. It had to start something like this:

Theorem. The square of the hypotenuse of a ...

Construction. Let *ABC* be a right triangle with $\angle A = 90°$. Let *D* be the foot of the perpendicular from ...

Proof.

$$\triangle XYZ = \triangle PQR \text{ (three equal sides)}$$
$$\therefore \angle XYZ = \angle PQR$$

$$\vdots$$

QED.

What generations of schoolchildren took away from this kind of thing was:

1. An ingrained idea that proofs are incomprehensible.
2. The knowledge that $\therefore$ means "therefore" (and it's real cool to scatter $\therefore$'s all over every $\therefore$ thing you write $\therefore$).
3. That you should always end a proof with QED.

At university they promptly get the latter two items beaten out of them, learning the hard way that in place of the sacred three dots you should use the word "therefore," and that the proper way to end a proof is to stop in what looks like the middle and draw the "Halmos symbol" □. They remain convinced that proofs are incomprehensible, though, so some of what they were taught at school does survive.

What Is a Proof?

Mathematicians spend an awful lot of time trying to teach their students to understand, and generate, proofs. Over three or four years the students learn how to produce a semblance of understanding—like someone who can write wonderful essays on the thematic structure of Romeo and Juliet until you ask them "hang on, just what is a balcony?" A lot of this is the fault of their teachers. When we try to tell them what a proof is, then just like those old-fashioned schoolteachers with their ∴ and QED, we home straight in on form, not substance.

In order to avoid being nasty to people who can't defend themselves, let me quote from one of my own books [9].

Definition. A *proof* of the statement $P \Rightarrow Q$ (where P and Q may be statements or predicates) given the statements $H_1, \ldots, H_r$, consists of a finite number of lines

$$L_1 = P$$
$$L_2 = \ldots$$
$$\vdots$$
$$L_n = Q$$

where each line $L_m (2 \leq m \leq n)$ is either a hypothesis $H_s (1 \leq s \leq r)$ or a statement or a predicate such that $(L_1 \& \ldots \& L_{m-1}) \Rightarrow L_m$ is a true statement.

Here $\Rightarrow$ is the logical symbol for "implies" and & is the logical symbol for "and." So my definition tells us that a proof is a sequence of statements, starting from the hypotheses and ending with the conclusion, in such a manner that each statement is a logical consequence of the previous ones.

It is this concept of proof that "structured proofs" buy into. Understanding becomes something that can be ticked off on a chart by a bureaucrat. "Why is statement 3.1.1.1.7 valid?" "Because statement 2.4.3.3.6 states that statement 1.9.3.3.2 implies statement 3.1.1.1.7, and statement 2.4.3.3.7 states that statement 1.9.3.3.2 is valid." I don't want that kind of understanding from my students. ("What does statement 2.4.3.3.7 tell us about rational numbers?" "Sorry, what do you mean, 'rational'?") I want answers more along the lines "Look, it all follows from uniqueness of prime factorization and the parity of the power of 2, right?"

Structured Shakespeare

    (II) ACT (II) SCENE.
      <1> Orchard belonging to Capulet.

themselves indelibly on your mind the very first time you heard them ...

In our culture, the stories that imprint most deeply are nursery tales. "Once upon a time there were Three Billy-Goats Gruff ... " The stories of Cinderella, of Rumpelstiltskin, of Puss in Boots, of Chicken Little who thought the sky was falling down ... Let me sidestep, for the moment, the question of why these stories imprint so indelibly, which involves deep-seated psychological resonances as well as a compulsive story-line. Parents who are aware of the hidden meanings behind many nursery tales will discover that many of them are suitable only for adult viewing after the ten o'clock watershed, and totally inappropriate for tender minds. Instead, let me focus on their lighter aspects, taking as my text the tale of *The Three Little Pigs* [10]. I'm sure that readers who are unfamiliar with the tale will be able to deduce its outline from the version that I am about to give, which is the story of the Pigs as it might be told by a mathematician writing for a conventional research journal. It is a deliberately frivolous example with an entirely serious purpose. I include it to demonstrate just how thoroughly that style destroys any prospect of following the story-line.

The tale's journal title would probably be "On Certain Aspects of Triples of Infinitesimal Pigs," but in the interests of obscurity I have opted for a more symbolic one:

$$\text{Let } p_j = \{\text{pigs}\} \cap \{\text{little}\}, \quad j = 1, 2, 3.$$

The entire story of The Three Little Pigs can now be told in graphical and symbolic form, as shown in the Appendix.

It may be worth making a few features of the mathematical "story" explicit. The simplest comparison between that and the nursery tale (obliterated by my scissors-and-paste editing) is by area of page used. When you see large expanses of mathematics and no pictures, the nursery tale is the more economical; when you see large expanses of white paper with few symbols in them, the nursery tale is the less economical.

Once you've got over the symbolism, the most notable feature of the mathematics is the early appearance of four pages, unrelieved by pictures, consisting entirely of definitions. This technique bears fruit later—where, for example, ten lines of nursery version text reduces to the symbols n_3 and $\wp_3$. But it makes the story very obscure.

I included this piece of porcine foolery as a gentle way of suggesting that, as a profession, we mathematicians generally tell our stories exceptionally badly. The symbolic version of The Three Little Pigs is virtually unreadable, even to a trained mathematician. The story-line is almost totally obscured — not by the symbols, which any mathematician can read and interpret easily, but by the style. It would not be so hard to rewrite the symbols so that the story-line once more became apparent to the casual eye. Merely introducing the symbols when they are needed, rather than en masse, would go a long way. Good writers of mathematical papers make use of a thousand more or less subtle techniques of this kind to ensure that the mathematically literate reader can follow their story. Bad writers — and there are many — simply plough ahead into the symbolism, and expect the reader to reconstruct their meaning from whatever clues they offer. New authors copy a random selection of earlier ones, who are mostly bad writers, and the level of clarity goes into a slow but inevitable decline. Goldwasser's file cards beckon.

"Structured proofs," of themselves, simply do not address this problem: they focus on format, when they should be focusing on style. Everybody who teaches computer programming knows that a program with perfect syntax, which carries out its allotted task efficiently and accurately, may be totally unreadable because of poor style. "Structured proofs" attempt to import programming syntax into mathematics, but they will only fulfill their promise if a great deal more attention is paid to programming style.

The Self-Effacing Fox

Even elegant style can obscure meaning, rather than enhance it. This is not a new problem. Arguably the greatest mathematician who ever lived was Carl Friedrich Gauss. He was the son of a farm laborer, mathematically precocious, and in later life relied — as did many of the subject's leading figures — on the sponsorship of the nobility. Gauss was a complex figure, and it would be unfair to call him a snob; but in certain respects, he displayed an unfortunate kind of academic snobbery that made his ideas seem much more mysterious than they often were. He defended this habit, saying that "When a fine building is finished, the scaffolding should no longer be visible." This is mathematics-as-art, or rather mathematics-as-artifact: the view that only the finished product is of interest, not the process that created it. Anyone whose focus is on producing mathematicians, rather than mathematics, can see the pitfalls. Another first-rate mathematician, Niels Henrik Abel, lamented that Gauss was "like the fox, who effaces his tracks in the sand with his tail." Carl Gustav Jacob Jacobi called his proofs "stark and frozen ... so that one must first thaw them out."

One of Gauss's pet bugbears was the so-called "fundamental theorem of algebra," which states that every polynomial equation has at least one solution in complex numbers. During his lifetime he produced several different proofs of this basic theorem — all of the frozen variety. Mathematicians can read a very interesting instance in Meschowski [7], where the fox has used his tail so effectively that not only is the scaffolding invisible: the entire edifice appears to float unsupported in mid-air. For the rest, let me offer a dialogue that is in much the same spirit.

GAUSS: I shall prove that every polynomial equation has at least one solution in complex numbers.

ME: Fine. Go ahead.

GAUSS: Suppose, for a contradiction, that there exists some polynomial equation $f(z) = 0$ that has no solution z in complex numbers. Then $f(z)$ is never zero.

ME: Absolutely.

GAUSS: Then, as every student of complex analysis knows, it follows that every glubbinzapper must be a phofflbont. Correct?

ME: I'm sure you're right.

GAUSS: I am, it's in all the textbooks — look it up if you're worried.

ME: Let's take it as given.

GAUSS: OK. I'm now going to tell you how to start from this supposed polynomial $f(z)$ that is never zero, and make a glubbinzapper that isn't a phofflbont. Then that will be a contradiction, and my theorem will be proved.

ME: Fair enough. Let's see it, then.

GAUSS: Well, it's a bit complicated ... I'll start by making two snurglewungs and two thimpets, like this... [*He goes on for several pages ...*] So now I've got myself a glubbinzapper, right?

ME: Reckon you have, yes. Can we just check that argument on page three once more, to be on the safe side? [*We do. It holds up fine.*] Yup, I'm convinced.

GAUSS: Great. But my glubbinzapper can't possibly be a phofflbont. You see, if it was a phofflbont, it would have pink snodphuns, but you can see from this routine ten-page calculation that actually one of its snodphuns is purple.

ME: I'll have to check, but I'm sure you're right. Well, that's amazing. Er ...

GAUSS: [*Irritated.*] Er what?

ME: Er—isn't there a more, well, straightforward way to prove it?

GAUSS: When a fine building is finished, the scaffolding should no longer be visible.

What I Tell You Three Times

Once we start talking narrative, rather than formal logic or programming syntax, we enter a realm in which reductionistically trained mathematicians feel distinctly uneasy. Unlike the crystalline rigor of the propositional calculus or axiomatic set theory, the whole business feels new-agey and vague. The effectiveness of narrative draws on millions of years of human "extelligence" (see Stewart and Cohen [8]), or cultural capital. It employs trickery that lies far deeper in our psyche than humanity's recently evolved surface smudge of rationality.

For example, it is curious how many nursery tales involve threefold repetition. The Three Billy-Goats Gruff, Goldilocks and the Three Bears, the Three Little Pigs ... In *The Hunting of the Snark* [4] Lewis Carroll said "What I tell you three times is true." Though, as David Fowler recently pointed out to me, he says this only twice in the book ...

Why three?

We can imagine a culture in which that number was different. Mathematicians might care to set up The N Little Pigs for any N, perhaps including the case of N an infinite ordinal. However, no sane parent would read The 101 Little Pigs to their children, let alone The ω Little Pigs. Except to put them to sleep, a time-honored task—the Canadian analogue of the song The Twelve Days of Christmas has 365 verses, one for every day of the year.

On the other hand, The Two Little Pigs really wouldn't work. Pig One gets eaten, Pig Two doesn't—what's the Big Deal?

Pig One sets a pattern. But it might be any of a dozen patterns, and we can't be sure which. Maybe the next protagonist will be a sheep. Pig Two confirms the pattern, and thereby determines a progression—which our fertile pattern-seeking minds immediately extrapolate in the obvious direction. We know at this stage that if the progression continues, Pig Three will get his house blown down too, and be eaten by the wolf. Narrative economy now dictates that it is Pig Three that breaks the pattern—thereby making the point of the story, which is, among other things, that the wise person learns from the failures of others.

Exactly the same thing happens in Goldilocks and in Billy-Goats Gruff.

There is one problem with this explanation: if the traditional nursery tale had happened to be The Four Little Pigs, I could have explained the numerology just as convincingly. ("First two set up pattern, third confirms it, fourth breaks it.") Compare Chicken Licken and the sky that was falling down, which continues the repetitive sequence for seven or eight steps. In fact, it may be that "three" has the effect that it does because we are exposed to a stream of nursery tales that have that structure—and not the other way round.

This is the darker meaning of the pigs. Our outlook on life can be deeply affected by the stories that we are told. Stories touch a primal part of our psyche, an irrational part. But psychologists now tell us that without emotional underpinnings, the rational part of our mind doesn't work. It seems that we can only be rational about things if we have an emotional commitment to such a recently evolved technique as rationality.

This is probably why academics argue so violently.

I don't think I could get very emotional about a structured proof, however elegant. But when I can really *feel* the power of a mathematical story-line, something happens in my mind that I can never forget. Goldwasser's attempts to automate journalism produce the outward semblance of news, but they fall apart because they don't tell a coherent story. I'd rather we improved the *storytelling* of proofs, instead of dissecting them into bits that can be placed in stacks of file cards and sorted into order.

References

[1] Garrett Birkhoff and Saunders MacLane. *A Survey of Modern Algebra*. Macmillan, New York: 1953.

[2] Michael Frayn. *The Tin Men*. Collins, London: 1965.

[3] Jack Cohen and Ian Stewart. *The Collapse of Chaos*. Viking, New York: 1994.

[4] M. Gardner. *The Annotated Snark*. Penguin, Harmondsworth: 1962.

[5] John L. Kelley. *General Topology*. Van Nostrand, Princeton: 1955.

[6] Leslie Lamport. "How to write a proof." *Amer. Math. Monthly*, 102 (1995), 600–608.

[7] Herbert Meschowski. *Ways of Thought of Great Mathematicians*. Holden-Day, San Francisco: 1964.

[8] Ian Stewart and Jack Cohen. *Figments of Reality*. Cambridge University Press, Cambridge: 1997.

[9] Ian Stewart and David Tall. *The Foundations of Mathematics*. Oxford University Press, Oxford: 1977.

[10] Joan Stimson (illustrator Steve Smallman). *The Three Little Pigs*. Ladybird Books, Loughborough, UK: 1993.

Appendix—The Three Little Pigs

Define f: $\mathcal{P} \to \mathcal{H}$ such that $f(p_1) \in \{\text{straw}\}$, $f(p_2) \in \{\text{sticks}\}$, $f(p_3) \in \{\text{bricks}\}$.

Let $h_j = f(p_j)$, $j = 1, 2, 3$.

Suppose that

$$\exists! \; W \in \{\text{wolves}\} \cap \{\text{big}\} \cap \{\text{bad}\}.$$

Let q_1 = "I'll huff", let q_2 = "I'll puff", let q_3 = "I'll blow your house down".

Let r_1 = "he huffed", let r_2 = "he puffed", let r_3 = "he blew the house down".

Let n_j = "'"now W won't catch and eat *me*," said p_j".

Let L = "let me in, pig, where pig $< \varepsilon \ll 1$."

Let O = "no, no, by the hair of my chinny chin chin, I will *not* let you in!"

Let R_j = "and p_j ran to h_{j+1}", $j = 1, 2$.

Let $\Omega = \{h \in \mathcal{H} : h \text{ would not fall down}\}$.

Let $\Delta = \{\text{doors}\}$ and define the relation $\approx$ ('knocks on') $\subset \{W\} \times \Delta$.

Let $\Pi = \{pots\}$, $\Phi = \{fires\}$, $\Xi = \{chimneys\}$, $\Re = \{roofs\}$.

Choose fixed but arbitrary $\pi_0 \in \Pi$, $\varphi_0 \in \Phi$, $\xi_0 \in \Xi$, such that $\pi_0, \varphi_0, \xi_0 \in p_2$.

For $j = 1, 2$ let $Q_j := [$ "L" said W. "O" replied p_j. "Then", said W, "q_1 & q_2 & q_3" $\therefore r_1$ & r_2 & r_3. So R_j.]

Choose d_j $\Delta l h_j$, $j = 1,2,3$, sufficiently small that W $> d_j$ for $j = 1, 2, 3$.

Define $\beta^* =$ "built".

Let $\wp_j :=$ "$p_j \beta^* h_j$".

Let $A_j := $ W 8 p_j

Let $p_j \in$ {pigs} $\cap$ {little},
$j = 1, 2, 3$.
Let $\mathcal{P} = \{p_j \mid 1 \leq j \leq 3,$
$j \in \mathbb{N}\}$.
Let $\mathcal{H} =$ {houses},
assumed for simplicity to
be compact Hausdorff.

(A routine argument
shows that Hausdorff can
be relaxed to T_1 provided
the Riemann Hypothesis
holds.)

Theorem 3.pigs
(WOLF'S DILEMMA)

$$\exists \delta > 0, \delta \in \{\text{days}\} :$$

$\wp 1$

$n1$

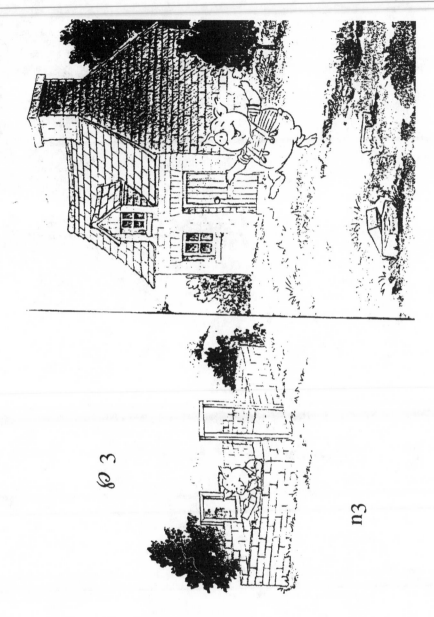

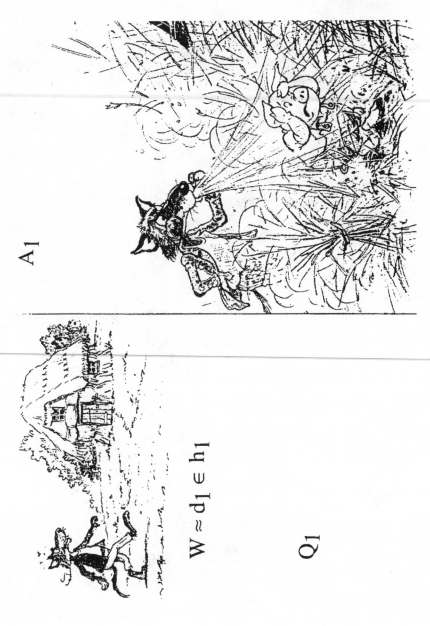

A₁

$W \approx d_1 \in h_1$

Q₁

$$W \approx d_2 \in h_2$$

Q2

A2

$W \approx d_3 \in h_3$

"L" said W. "O" replied p3. "Then", said W, "q1 & q2 & q3".

∴ r1 & r2 & r1 & r2 & r1 & r2 & r1 & r2 & r1 & r2 &... but h3 ∈ Ω.

Exercise for the reader

!*@!!!%%!!!!!

A routine but tedious calculation shows that p3

$\Rightarrow \pi_0/\varphi_0$, where

$\varphi_0 \in \Xi|h_3$.

$\exists$ surjection $\sigma: W \to \Re|h_3$.

$W \downarrow \chi \in \Xi|h_3$.

Hence, by continuity,
$W \downarrow \pi_0$.

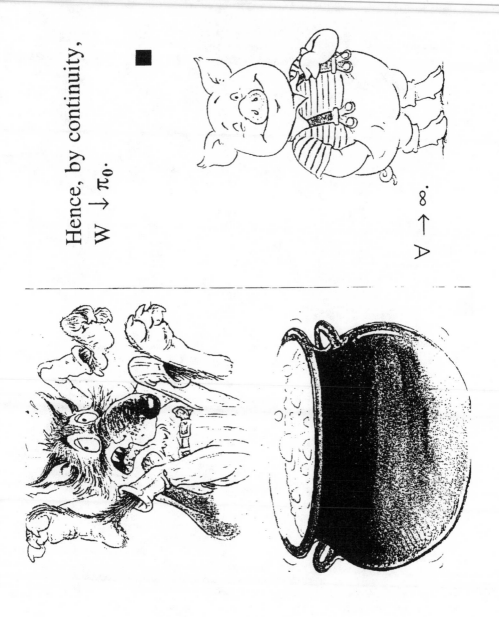

$A \rightarrow \infty$.

Contributors

John D. Barrow — Astronomy Research Centre, University of Sussex, Brighton BN1 9QH, UK (email: j.d.barrow@sussex.ac.uk)

Greg Bear — 506 Lakeview Road, Alderwood Manor, WA 98037 (email: grbear@ix.netcom.com)

Gregory A. Benford — Department of Physics & Astronomy, University of California, Irvine, CA 92717 (email: molsen@uci.edu)

John L. Casti — Santa Fe Institute, 1399 Hyde Park Road, Santa Fe, NM 87501 (email: casti@santafe.edu)

Jack Cohen — Mathematics Institute, University of Warwick, Coventry CV4 7AL, UK (email: kc@dna.bio.warwick.ac.uk)

Per Johansson — Human Ecology Division, Department of European Ethnology, University of Lund, S–223 62 Lund, Sweden (email: per.johansson@humecol.lu.se)

Kjell Jonsson — Department of History of Science and Ideas, Umeå University, S–901 87 Umeå, Sweden (email: kjell.jonsson@idehist.umu.se)

Anders Karlqvist — Swedish Polar Research Secretariat, Royal Swedish Academy of Sciences, Box 10005, S–104 05 Stockholm, Sweden (email: anders@polar.kva.se)

Paul J. McAuley — 83a De Beauvior Road, London N1 4EL, UK (email: mcauley@omegacom.demon.co.uk)

Larry Niven — 3961 Vanalden Avenue, Tarzana, CA 91356 (email: fithp@aol.com)

Ian Stewart — Mathematics Institute, University of Warwick, Coventry CV4 7AL, UK (email: ins@maths.warwick.ac.uk)

Index